# the Burned-Out Blogger's Guide to PR

## Jason Kincaid

# Contents

*For Jessica, Chloe, Cassidy, and Mr. Cody*

Good dogs.

# Introduction

I wrote this for two reasons.

The first is why I started.

Revenge.

For four years as a writer at *TechCrunch* the PR industry made me miserable. There are great people in PR, but their numbers are paltry next to the hordes of brown-nosed hucksters spewing nags and deception, desperately trying to justify their hefty paychecks. I figured if I wrote this, maybe a few of the real stinkers — the sort who can be supplanted by a book — would be forced to go off and do something other than drive reporters insane.

The second is why I finished.

Right now there are thousands of people working on startups.

Many have ostracized themselves from everything they once cared about, are leading the grim lives they've been told to endure if they want to succeed like their heroes. The odds are stacked against them. Most will fail. The fact that some will shoot themselves in the foot over something so trivial as a press embargo, or by hiring a bad PR rep, drives me nuts. More than the brown-nosers ever did.

Jason Kincaid
February 2014

# A Note on Syntax

It's been a while since I was a tech reporter, but for the sake of clarity this book is written as if I still am one. A lot of it is self-flagellation, anyway.

I make broad generalizations about reporters and founders. There are superlative exceptions in both camps.

As with most advice, this book stems from my experience. My goal is honesty. I haven't polled a bunch of reporters to see if they agree, because reporters never agree on anything.

You'll notice I favor the word "reporters" over "journalists". Journalist is a word that gets people riled up, and this isn't the place for a debate over what constitutes what.

Let's get to it.

# Unembargoed Homily

The tragedy of tech reporters is that they inevitably grow numb to the spark of innovation that got them involved in the first place. Products designed to dazzle and disrupt — some, the fruits of years of work — take on the homogenous diversity of a Taco Bell, an underwhelming array of innovation made from the same six ingredients.

Cynicism buds. Startup names congeal into an amorphous blob of meaninglessness. The cacophony of manufactured excitement drowns out nearly everything genuine. Occasionally something special reminds them why they're there — but it isn't long before another slimeball PR exec cold-calls at 8 PM on a Thursday and their hearts cool again.

Good reporters trudge forward and never stop loving good stories: shining the spotlight on something that deserves it is good for the soul and can pay career dividends. But the endless streams of emails and handshakes carve out grooves that dictate how they find these stories, using filters deliberate and subconscious, often rewarding the already-successful.

In other words, you've got your work cut out for you.

There is no secret to getting press. Even the most press-savvy entrepreneurs and PR executives have horror stories of botched launches and ignored pitches.

Because while it's nice to think that reporters have structured and fair processes for deciding who is blessed with press coverage, you are at the mercy of the black box that is the reporter's mind. They're well-versed in explaining why they chose to write one story over another, but this deliberation is roughly as scientific as how they decided what to order for lunch. Which isn't to say they're wrong — instinct is everything in this business — but there is no rubric you can incredulously point to when they say, "I'll pass on this one."

Of course every publication has guidelines around what is worth covering, and all writers want to write interesting stuff. But the reality of the newsroom simply doesn't allow for much more than gut instinct. There are heaps of pitches to sift through, and time spent reading pitches is time not spent writing and reporting. Editors may be involved in the process, but they're mostly concerned with approving and improving stories — they don't hear about every pitch a reporter decides not to cover.

Maybe you send your pitch and hear crickets. Or maybe your favorite reporter responds immediately, and they're super-psyched because you built their dream app. But just as they start writing their article the phone rings — Facebook, breaking news — so they forward it to one of their colleagues, and he's not feeling it, so here we are a week later and neither of them is giving you the time of day.

It's more of a crapshoot than anyone would like. But there are pitfalls you can avoid that'd otherwise assure certain failure — and a few tricks that may get Lady Luck to smile down on you.

Or at least, maybe she'll read your email.

# What is PR?

PR stands for Public Relations. You knew that. They get you press, they spin the bad stuff. You may have a sense there's something scummy about the whole thing (you are not wrong). But you probably don't realize how deeply and disturbingly influential the PR industry is.

Every day, hordes of PR people* inundate reporters with pitches. Some of them are hawking new apps, others trend pieces that just happen to paint their client in a positive light, or competitors in a bad one. Their numbers are legion: for every

---

* Ironically, and perhaps not accidentally, there is no generic single-word term for "PR people" (as there are, say, accountants or reporters). Some use the term PRs, which makes no sense, though maybe it's better than the glossy "PR Professional". I'll stick with "PR people".

reporter in the United States there are four working in PR — and they get paid better, to boot[†].

They are crafty, good at what they do. Too good. Many of the news stories you read every day originate not in the reporter's mind, but from the PR person who seeded it there. Reputable writers make the story *theirs*, critiquing and analyzing and so on, but they have deadlines to deal with — and the PR people know it.

Oh, what tricks they have.

Exhibit A: the press event, wherein a company beckons to reporters with the nebulous promise that it has "something exciting to share". The reporters are shepherded into a room where they are given meticulously crafted demos and free, catered food — and a fraction of the time they'd need to write anything remotely comprehensive. Much better for them to write their stories with the quips and tasty lox fresh in mind, before their skepticism gets a chance to fester.

But even this effort is often unnecessary. PR people wield tremendous influence over members of the press by simply giving them what they crave: pitches that are trivially easy to convert to stories, a buffet of quotes and stats and screenshots that need only a cursory round of paraphrasing to qualify as an article. Too often, they oblige.

PR people are even involved in stories where their influence is intentionally obfuscated. Every day, representatives from tech companies call and talk to reporters "Off the Record" and "On Background", planting ideas and corrections under the mutual agreement that the reporter will not publicly state that the

---

† "The U.S. boasts 4.6 public-relations specialists for every reporter or correspondent. Those PR pros earn 40% more a year on average than journalists." - The Wall Street Journal[1]

conversation occurred[+]. It's not quite as sinister as it sounds —
often, it means that reporters are less wrong than they would be
otherwise — but the takeaway is this: PR is everywhere, even
where it ostensibly isn't.

I'll leave further handwringing over this relationship to the
experts. Just remember: as a rule, reporters do not like PR people. But if all of them went up in smoke, there'd be an awful lot of
slow news days.

---

[+] You can still find clues. Keep an eye out for My understanding..., I believe..., and
even My hunch..., particularly when a usually well-informed reporter makes an asser-
tion that doesn't have much evidence to support it.

# What Does a PR Person
# Actually Do?

First comes the behind-the-scenes stuff. They'll meet with you to learn about your background and the product, then help steer you toward your Story, which is what everything else revolves around. They'll outline long-term strategy, map out milestones that can potentially be used as pitch-pieces, and help choose which reporters to contact.

Depending on the firm and how much you're paying, they may offer a variety of other services like "media training" to teach you how to talk to reporters, and event-planning for those lox-laden launches; some will even write your tweets for you.

This is all invisible to the reporter, who sees the PR person in a different context — namely, as the intermediary between themselves and founders. In this capacity, PR people do six things.

### 1. Contact the reporter with the initial pitch.

This is often an email, though sometimes the PR person will mention an upcoming story at a cocktail party, or over the phone. This step is crucial; if the reporter doesn't read (and understand) the pitch, it is dead in the water. Good PR people will go to great lengths to avoid this scenario: multiple emails, phone calls, and assorted other pestering. But — and here's the key — they don't go so far that the reporter hates them or the company they're representing (or, at least, any disdain they inspire has faded by the next time they come calling).

*I know how to write emails, you may be thinking.*

### 2. Answer questions.

Say you've sent me, the reporter, a pitch that I'm interested in. Usually there's something I'm going to ask about: Are you disclosing the terms of this deal? How does this feature work? How come you don't have this thing that I, in my infinite wisdom, think you should have?

Since the PR person sent over the pitch in the first place, I'll send my questions to them, and they'll help you come up with some answers (and if I send them directly to you, you'll be instructed to run your answers past the PR person before responding).

In other words, they're a filter. Their goal is to make sure everything you say is quotable (because it may be quoted), succinct (because my attention-span is minuscule), and chipper (or, in the case of bad news, appropriately somber). If your answers are complicated, they'll translate them into words I understand.

*I know how to respond to questions, you may be thinking.*

### 3. Handle logistics.

The PR person will try to coordinate an interview over the phone or in-person; if it's over the phone they'll set up the conference call and likely be on the line, if it's in-person they'll be sitting there with you. They'll manage the emails about embargoes, send resources relevant to your pitch, and give me gentle reminders as your launch approaches.

*Logistics, schomistics.*

### 4. Save you from yourself.

Some founders are bad at explaining things under pressure (and interviews with the press can feel intensely high-pressure). Some will dodge innocuous questions and respond with meaningless drivel. Some come off as arrogant jerks. Some badly need a Red Bull IV.

A good PR person will deflect the 'gotcha' questions, jump on grenades, subtly interrupt your impending bouts of verbal diarrhea, and do other things you won't even realize are saving your butt.

*Don't assume you don't need this.*

### 5. Buy you time.

It's 8pm. You're eating dinner — lunch, technically — after the longest day of your life. Your startup of eight had to let three go today, not enough money to justify the personnel. One of them had tears on his cheeks as he walked out. He'd moved seven hundred miles to be here. You're stressed and tired and just want to go to sleep but have thirty more emails to get through.

Your phone rings.

It's me, some reporter whose name you vaguely recognize. I've got a source telling me the company is imploding, I'm publishing soon unless you say something that makes me think otherwise, and I'm calling to see whether you have any comment.

Your mind fills with rage, dread.

"Who — who told you that?" you stammer.

I'm not surprised you asked, but I'm not telling. "A source." A beat later and I'm asking whether it's true that your core product is selling short of expectations and you have only three months of runway left.

There's a good chance this doesn't end well.

Now, let's picture an alternate universe, where:

I call you. You don't pick up. I call your PR rep who says she'll check in with you and get back to me immediately.

She calls you, shoots me an email saying I'll have a statement in half an hour, and spends the next thirty minutes drafting something that acknowledges the layoffs but explains them away: while it's true that a portion of the product wasn't working out (hence, the three let go), this is really a move about doubling down on the features your users love (half of whom come back daily).

I know it's pure spin, but I include your statement in the post anyway.

*How about I don't pick up for an unknown number on bad days, you may be thinking.*

### 6. Maintain a relationship.

Remember: your first hurdle is getting someone to pay attention to your pitch, and, trite as it sounds, having a recognizable name in the 'From' field goes a long way. Even better for that familiar name to carry a whiff of credibility that extends to the contents of the message, so that the voice in my head says, "Make sure to read this one."

You can accomplish this feat by asking a big-name investor or entrepreneur for an intro, but there's a good chance you don't know any. Alternatively you can buy yourself some attention, because, you guessed it: reporters recognize PR people names. Some of them, at least.

PR people and reporters see each other at parties and media events all the time. They've had their spats, smoothed them over with drinks, usually they aren't friends per se, but they have established relationships — the benefits of which extend beyond recognizing names in inboxes.

If a PR person has a good rapport with me, then I'm more likely to give them the benefit of the doubt in clutch situations, when, say, they ask me to hold a story for half an hour while they draft a statement.

Of the six points listed here, this is the most important.

*You can't magically make your name leap from a reporter's inbox.*

# Why Not to Get a PR Person

Clearly there are benefits. Now let's talk about why you wouldn't want one.

**Big Reason 1: The timing is off.**

Hiring a PR person long before you have anything substantial to announce will dent your bank account and can hurt your reputation with the press (a lot of PR people would rather spam reporters with non-news than point out that your money is better spent elsewhere).

Also: hiring a PR firm requires a time commitment. In the long run they can save you a lot of headaches, but the success of your PR person is predicated on what you give them to work with. PR relationships often fizzle because the client gets caught up in their countless other battles.

**Big Reason 2: PR is expensive.**

You're looking at a retainer of $5,000 to $10,000 and up. That's per month. For the *really* good, cream-of-the-crop folks, you're forking over equity. There are firms and independent PR people who charge less, but it won't be pocket change — and the less you pay, the more you should ask yourself why you're getting such a deal.

Good PR people can command high rates for several reasons: they know what they're doing, they have relationships with reporters, and they aren't in the habit of hawking junk pitches (because they can afford to turn away the boring companies). It's a simple recipe, but it's one they've perfected over many years, particularly when it comes to building those relationships.

I'll read their emails because I know they aren't going to waste my time. In fact, great PR people send over compelling stories so consistently that even the borderline stuff I might shrug off otherwise seems just a tad more interesting.

You probably can't afford one of the top firms. Not the end of the world. Let's slide a little further down the totem pole.

**Big Reason 3: A mediocre PR person is a waste of money.**

Here, you'll find plenty of options — still pricey, but less so — and there are some great people working at them. But they're less selective about the clients they take on, which means the same Account Executive responsible for your company may be moonlighting as a huckster for some social network wannabe, a total snoozer, so their name isn't exactly a shining star in my inbox.

You also face an increased risk of getting someone who has an undeveloped sense for what makes a good story, or a fragile grasp

## PR People Success Rates
(Cost vs Batting Average)

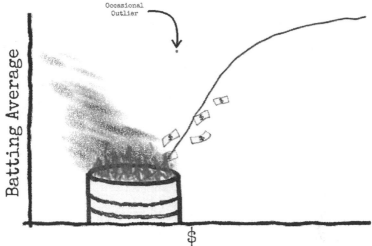

of your product (this often manifests as an understanding of the basic functionality, but an inability to discuss any specifics without having to check with the founders). But still, there's plenty of talent to be found.

Now let's venture down that totem a bit further. The prices are lower here — finally — but something unintuitive is starting to happen. Where before, at the high end, you could expect a reasonable chance of converting your dollars spent into press coverage, suddenly your batting average is dropping precipitously.

The reasons for this are best demonstrated with an example.

Let's say you've hired a bright but inexperienced PR guy named Chris. Obviously it'd have been nice to hire a veteran, but you're a startup — the list of "nice to haves" is endless — and Chris seems to know what he's doing.

Things go great at first. He helps chisel the company's story, writes your email pitch, and, as Launch Day approaches, sends emails to the half-dozen reporters you've identified as potential candidates. The email looks like this:

Hey Jason,

Loved your article on Perfecticle last week. Wanted to chime in that I'm working with a company launching this week that's doing something related and very, very cool.

They're called PixelYak. They're launching Thursday and we're asking for an embargo of 9AM PT. Please let me know if you agree to the embargo as I'd love to send over more details and set up a call.

And so on.

Now, this isn't a terrible start (there are things I'll gripe about later, but many pitches are worse). The problem is that there's a good chance I didn't make it past the first line.

I don't know who Chris is. Maybe I've seen his name before, but he belongs to the bucket of literally hundreds of other names that sound faintly familiar, another faceless member of the PR army that's been storming my inbox for years. He hasn't done anything wrong, but I have no relationship with him. Generic is probably the right word.

In this example it would have been better for you, the founder, to have sent the email. It's genuine — I know you're not a mercenary paid to believe in your product — and our exchanges don't have to pass through a middle-man, which takes longer and can introduce confusion. Plus, you'd have saved a bundle.

Granted, Chris did help you, he gave you advice on how to

frame the company and wrote most of the pitch itself — but that doesn't do you much good if no one is paying attention to it. And that's the problem with having a mediocre PR person: they're missing the relationships, and those are really what you're paying for[*].

Incidentally, there is no good way to discern the competence of PR people from their job titles. There's a hierarchy to it — fresh-faced Account Coordinators beget more-seasoned Account Executives beget Senior Communications Grand Poobahs — but firms are inconsistent and incentivized toward title inflation. This isn't to say you should shy away from Account Executives — some of them are great — but that these titles sound elite by design.

**Big Reason 4: A bad PR person will hurt you.**

A mediocre PR person runs the risk of getting you no press, but at least they leave you to try another day. A bad PR person can be disastrous.

Your PR person serves as your resident ambassador in the Nation of Disgruntled Reporters. They sometimes make judgement calls that they believe are in your best interest without checking with you first, and, because they have presumably been trained and seen it all before, can manage this task just fine.

But some of them are a little tone deaf. Some are overzealous to the point of being obnoxious and intrusive. Some just aren't that bright. These aren't bad people (with a few exceptions), but they're not the right people to be representing you. Their poor decisions can have a deleterious impact on your company.

---

* Of course there are outliers. Some of the best PR people I've worked with were newcomers on their way up, and some high-priced execs aren't worth it. The point is, if you're going to pay for it, don't skimp.

From the reporter's end, symptoms of a bad PR person include: incessant emails and phone calls, flakey embargo coordination (or worse, broken embargoes), whining, and lies. This stuff wouldn't be so bad if it only happened occasionally — and these folks count on their little disturbances being the exception — but it all adds up to an assault on my mental health.

And the buck doesn't stop with them. Doesn't matter that you had nothing to do with coordinating the train-wreck of an embargo, or that you had no idea that your Account Executive has been calling me every night this week. It reflects poorly on you, and reporters are a petty bunch.

# Why Do I Care About Press?

Let's take a step back.

Some founders have the idea that getting covered in a major publication is a golden ticket to fame and fortune. Others think that press is nice validation and good for filling bare office walls, but is mostly a distraction. Neither are right, but between the two, I'd bet on the latter.

You've probably heard of *Words With Friends*. The online Scrabble-esque game has since seen its heyday, but in 2010 it was a mega-hit that led its maker, Newtoy, to an acquisition by Zynga for $53 million.

What you may not know is that the company behind *Words* struggled in relative obscurity for years. In 2008 I wrote a post on

*TechCrunch* about their first game, aptly called *Chess With Friends*, which included innovative social mechanics that set it apart from the pack[*]. But the app's download numbers were modest; maybe my post helped a little, but nowhere near enough to sustain the company, which had to take on external projects to pay the bills.

Newtoy's followup, *Words With Friends*, launched in July 2009 and received coverage in several blogs including the iOS gaming site *Touch Arcade*, where it scored a coveted 5-star review. Despite this, the game's performance was middling. With a month of runway left, Newtoy's founders found their company on the brink of failure. So much for good press.

Then John Mayer happened.

Mayer, who had no affiliation with Newtoy, tweeted that *Words With Friends* "is the new Twitter". The game surged in the App Store (Newtoy's founders initially thought the spike in their 'downloads' graph was a fluke), and its inherent network effects took hold. *Words* became a household name, and the rest is history.

The moral of this story is not to go hunting for celebrity endorsements; the startup landscape is riddled with the tombs of companies who thought star-power was enough to sustain them. I'm just trying to paint the randomness of it all.

See, the press — even really positive press — is rarely going to move the needle in the long-term. It'll get you early adopters, but early adopters are fickle. The spike you see the day you launch

---

[*] The game featured asynchonous (i.e. turn-based) multiplayer. This sounds trivial now — few board game apps are released without it — but four months after the App Store's launch it was still rare. Part of the reason why: Apple had yet to launch push notifications, so you had to manually check the app to see if it was your turn.

will just serve to make your subsequent (lack of) growth look all the more depressing. You need to build something that people find worthwhile to get them to come back and tell their friends. Press will only help so much.

But still. Newtoy didn't change a thing, and *Words* flipped a bit and leapt to the top of the charts. Which means they *had* built something people found worthwhile. They just hadn't gotten lucky yet.

The way I see it, this is all a big lottery, and press is a good way to get more tickets.

Come launch day a post on *TechCrunch* will get you in front of thousands of people who otherwise never would've heard of you, then three days later you'll be wondering where everyone went. But even as your user-count fades — and that's the norm — there's a chance that something else is happening.

Maybe that blog post sets off a Rube Goldbergesque chain of events that leads to a tweet from John Mayer. Or maybe your John Mayer is a rich white guy investing other peoples' money, or maybe he's a she, and she thinks your nascent product may fit in nicely with her ambitions to rebuild her multi-billion dollar company. There are a lot of people who get paid to monitor the bleeding edge. You may not ever hear from them, but now you're on their radar.

Or maybe none of that happens.

A single post won't have a huge impact. But if you get the hang of the press cycle — something that is infinitely easier if you have a product worth writing about — then your odds improve dramatically.

Also: press is far from the only way to score these tickets. You can get more by going to networking events, by connecting with a great investor or advisor, by writing something insightful on your blog — and, most of all, by improving your product.

And if things don't work out, don't kick yourself over it so much. The only people consistently making money on the lotto are the ones selling tickets.

# Choose Your Destiny

Before we go any further, I want to make something clear: doing press yourself can be a time-sink and emotional burden, leaving you to simmer in angst and frustration over unreturned emails and shoddy reporting. And some people just aren't great at interacting with reporters; if you have to be bad at something, may you be so lucky for it to be a skill you can outsource.

But some of you — including the less-funded — may think this doesn't sound so bad. Founders certainly have to deal with worse headaches. Maybe there's someone on the team who's good at talking to people and has a way with words. Then again, maybe they have a hitherto-unseen tendency to play buzzword bingo as soon as a reporter interrupts their monologue. You'd be surprised.

You don't have to decide which group you're in yet, but I'm not going to make the decision for you.

The remainder of this book is written with the aim to help you navigate your way through the early press scenarios — pre-launch and post-launch — as if you were handling your PR yourself, covering terminology, tactics, and how to come up with compelling story ideas. Even if you ultimately decide it isn't for you, it's my hope that you'll be better equipped to identify a solid PR person.

But before we get into the nitty-gritty of pitch writing (god, don't make me talk about embargoes) there are a few things you need to understand about the people who make up the press.

# Under the Influence

Your average reporter is scary; a mob of them is terrifying.

Don't get me wrong — on an individual basis many of them are lovely, talented people who go to great lengths to ensure they are fair and accurate. But this is not the *average*.

Because while reporters wield great power, they are under the influence of an intoxicating booze of validation that has a profound impact on logic. Tweets, comments, page views and Likes: each article is given a score that amplifies the fact that hyperbole and drama are more exciting than nuance and normalcy. Reporters may not intentionally aim for sensationalism, but their black boxes are complex, and there is no denying that a post that performs better *feels* better. Sometimes the booze wins.

Its effects are readily apparent.

Behold, the reporter wondering aloud *\*hic\**, to an audience of millions, whether *This Precocious Startup Could Be The Next Facebook-Killer?*

But don't pop the champagne just yet, because someone just identified a minor bug in your app and *\*hic\** suddenly *Your Gaping Breach Is Putting Thousands At Risk.*

Now here come the rumors and anonymous sources — the high-proof stuff — as an *Ex-Employee Calls Your Latest Launch A "Massive Mistake"* (what's worse, *\*hic\** they gave you all of five minutes to comment before publishing).

And those are the nice reporters.

Some reporters aren't nice. Their mission is to report the news and if it hurts your feelings along the way, well — tell it to your shrink. They live and die for the story, to disrupt the status quo, and whatever other lofty mission they've conjured to rationalize their thirst for attention.

Anything they hear you say is fair game. They'll twist quotes and strip context in ways that would make reality TV producers jealous. They'll lie and tell half truths. They'll forget to mention they're a reporter. They'll badger you with questions when you're drunk. They'll call up past coworkers until they find someone willing to say something nasty about you, and refer to them as a "source who worked closely with you in the past."

Should you have the misfortune of crossing paths with one of these less honorable writers, you'll find that they will frame their article in such a way that everything they did was justified, and your options of recourse are effectively nonexistent.

Put another way, reporters can be tremendous assholes who believe they can invade privacy and shrug empathy because they belong to the high-and-mighty institution of the Press. There are some who frown on the techniques above, but there are others who don't — and none of this is against the law.

What I'm saying is, be careful.

# Don't Let the Reporter Be Wrong

This is harder than it sounds, and you will probably fail in this endeavor. But you must try.

Reporters are not a stupid bunch. Many of them are quite bright. But between the endless deluge of deadlines, the fear that their hackish rivals are about to beat them to the punch — and that boozy validation — they can't help but slip up. They misremember things, they make typos that cut your revenue by an order of magnitude, they reference news articles that were wrong six months ago and, lo and behold, are still wrong today.

If reporters were surgeons, the question of whether any sponges were left inside the patient would not be answered in terms of "if" but instead, "how many?" Your job is to make sure they are at least operating on the correct limb.

Your first weapon in this battle is email. But email itself is unwieldy, a blunt tool that can do as much harm as good. You must refine this email. You must make it impossible to misinterpret. You must weave Bolds and Bullets and Indents with the grace of a samurai, trimming away the unnecessary so that only the essential remains. Not simply that it can be found — but that it becomes *unmissable*.

Your second weapon is the interview. Typically conducted either in person or over the phone — I favor the latter, because it is faster and I don't like shaking hands — the interview is where you can engage in the dance of explaining what exactly you are doing. This, too, is an art.

A poor interviewee assumes the reporter fully grasps the words coming out of their mouth; a mistake, because the reporter was checking their email during the most crucial points.

A poor interviewee assumes the reporter is already familiar with their company's competitors. Granted, they do have some competitors in mind — but they are pitting you against heavyweights in an industry that you have nothing to do with.

A poor interviewee relies on common acronyms, making the reasonable assumption that the reporter knows what they mean. They are planning to look them up on Wikipedia afterward.

I jest*, but this is serious. During every step of your interaction you must be cognizant of things the reporter could misinterpret. The cost of having them get it wrong can be dire.

In Jurassic Park, researchers extract DNA from fossilized mosquitos to recreate long-extinct dinosaurs. During this process they run into a hiccup — some of the DNA has degraded — so

---

* Not really — I've done all of these.

they use the wonders of science to insert modern amphibian DNA to fill in the gaps (close enough, they figure). As a result of this slight genetic modification, the dinosaurs adapt in unforeseen ways, take over the park, and kill nearly everyone.

Reporters tend to fill in the gaps, too.

If a reporter does not fully grasp how your product works, they are not going to write an article that merely leaves out key details — they'd ask more questions if they felt like they didn't have the full story. The trouble is that sometimes they think they have the story when it is really another story entirely. Next thing you know you're reading about yourself attacking Google head-on, having cleared some technological hurdle that, as it happens, remains technically impossible.

Nature finds a way.

It's bad enough for a reporter to botch the story and disseminate incorrect information, but you also risk having them absolutely loathe you. Because while they're used to the vitriolic attacks of internet trolls, there is no worse feeling than getting something genuinely wrong. Maybe they'll attribute the screwup to sleep deprivation and feel badly about it, or maybe they'll blame you for giving a nebulous pitch and hate you forever. You don't want to take that chance.

No gaps.

# Relationships

One of the bizarre experiences a reporter goes through is the transition from mere mortal to object-of-desire. It is especially odd because — and I say this with affection — reporters are nerds. The only parties I was invited to in high school involved linking a dozen computers to play Quake 3 and exchange rhythmic footage of the human form.

And yet here I am, 22 years old, three weeks into my gig at *TechCrunch*, and I'm *cool*. Not just cool — I'm the friggin prom king. Every time the word "TechCrunch" comes out of my mouth it's as if I've just stabbed the person in front of me with a syringe of adrenaline:

"Oh wow! I love *TechCrunch*!" they say as they eye my name badge (never mind that I said it ten seconds ago).

"Jason, dude! Really love your stuff!"

Suddenly people are trekking across the room and back to introduce their friends, they're peppering me with questions — even the cocktail waiters are going out of their way to make sure I'm getting favorable access to the sliders and beer. I've become someone people get excited to talk to. My ego is growing faster than my untamed sideburns.

Then the emails start coming in.

> "Hey Jason, great meeting you last week at the StartupSiege dinner. Wanted to ping you about…"

> "Jason, bro! Have this awesome app my buddies and I have been building…"

And so on. None of this comes as a surprise — I'm not dumb, I know these people are interested in me for a reason — but I'm struck by the consistency. Even conversations that are friendly in nature, discussing things that have absolutely nothing to do with tech, always seem to result in one ask or another.

The parties get less fun. I go into every handshake subconsciously predicting what each person wants from me, how long they'll wait to ask, whether they'll act like they're just friends asking for favor, as if it hadn't been their intent all along.

Some people walk into a room and picture what everyone would look like naked. I see the formatting of their unsolicited email introductions.

And so we come to the topic of relationships, which, as I've said, play a vital role in determining who gets press coverage.

You may have decided that it'd be a good idea to make friends with some reporters. It's coming from someplace genuine: you're both in tech, seem to share a lot of the same interests, have mutual friends on Facebook, and sure, it'd be great to get your company written up at some point, but that's not what you're *really* after.

Except that's exactly what you and a thousand other people are thinking.

Smart PR people and entrepreneurs know better than to overtly "use" reporters. It's a much better investment to establish a long-term relationship, one founded in empathy and camaraderie, with a dash of pitching here and there.

A few manage to pull this off and are genuinely cool people, but many veer into an uncanny valley where the smiles are too wide, the compliments too calculated — and those are just the masquerades we spot. Eventually, reporters learn that friendship and Machiavellian patience look disturbingly similar. I've grown wary of both.

Look: networking is part of the game. I'm not saying that reporters need your hugs and pity, but want to explain why, in your interactions with us, you may find us aloof or even callous. More likely we'll be friendly but distant, our guard up out of habit. Everyone is trying to be our "real" friend.

This stuff works both ways. If you notice a reporter is being warmer toward you than you might expect, they may see you as a potential source. If you're comfortable with this you can use it to your advantage, but don't let them manipulate you. You don't owe them anything.

And go watch *Almost Famous*. Lester Bangs tells it better than I do.

# Party Time

A quick thing that ties back to relationships.

As you go to events and maybe house parties, you'll encounter reporters in situations that could be compromising. Maybe you're having a casual not-so-PC conversation, maybe you're all getting hammered or smoking pot or doing harder stuff. You're probably nervous, and for good reason — especially if no one has established that the guy with the Press badge is "cool".

So you turn to me and make a sorta-joke.

*"Hey man, this isn't gonna be on* TechCrunch *tomorrow, right?"*

And just like that I'm reminded I'm not really one of the gang, that at least some of the people around me view me with skepticism and fear, maybe even contempt — but are willing to feign friendliness so as to maintain their 'in' with the press.

It stings.

But if I'm in your shoes, and none of my buddies has vouched for the reporter, I ask. Probably not as a joke — because they know you're not really joking. Just a quick, unthreatening, "Hey, sorry for being paranoid, but this is all *Off the Record*, right?"

Some reporters probably think this isn't necessary, would rather they didn't need to feel that sting, but there are some jerks who spoil it for everyone.

∾

# The Record

Get excited: it's time for an aside on terminology.

Reporters run around with voice recorders and notepads and write down whatever they hear — then stick it on the internet for the masses to pass judgement. They'll quote your tweets, they'll quote your emails, they'll quote stuff they overhear at the next table. They're quoting machines.

Of course, no one wants to speak candidly or whisper secrets in their ears if it's just going to show up in the news an hour later. So reporters invented some special terms to put conversations into different buckets.

There's the default — *On the Record* — which means they can print whatever they hear and attribute it to whomever they heard it from. Unless the reporter affirms otherwise, you're *On the Record*. Pretty straightforward, no?

It's all downhill from here. Because while you may have heard the other key terms before — *On Background* and *Off the Record* — no one is really sure what they mean. Different reporters have different definitions, and there are a bunch of other terms, like "Not for Attribution" and "Deep Background", which, again, mean whatever they say it does.

The whole thing is asinine. Careers go up in flames when there are misunderstandings around attribution. But them's the breaks.

So: I'm going to describe the way I've understood these terms, but before I do that, let's lay down some ground rules.

> **Rule #1:** If you're talking to a reporter and don't want to see something printed, make sure you are both clear on what the terms are; don't assume you think they mean the same thing.

*Bad*

**YOU:** Can we go *On Background*?

**REPORTER:** Sure.

**YOU:** *\*Spills the beans, gets burned\**

*Good*

**YOU:** Can we go *On Background*?

**REPORTER:** Sure.

**YOU:** You can paraphrase this and say it came from 'someone close to the company', okay?

**REPORTER:** Can I say you're an executive there?

**YOU:** No, I'm too close to it. Can we stick with 'someone close to the company'?

**REPORTER:** Okay.

**Rule #2:** Make sure the reporter explicitly agrees to whatever 'bucket' you've asked to be in. Some people will casually throw in an "Off the Record" mid-sentence and immediately dish some dirt, and the reporter will a) get annoyed because they didn't want to go *Off the Record* or b) decide that they weren't really *Off the Record* because there was no consent, so now that dirt is about to get its time in the limelight.

This stuff is tricky. Reporters are usually okay with going *On Background* or, less often, *Off the Record*, but they'll get annoyed if you try to pinball between them, or if you try to share stuff anonymously that has no business being anonymous.

With that out of the way, here's how I've always used them.

**On the Record**

The reporter can quote you verbatim and attribute it to you. Assume everything — in person, email, text, whatever — is *On the Record* until the reporter agrees otherwise.

**REPORTER:** "I hear that Google is launching a new line of smart refrigerators."

**AARON DALE, Google Spokesperson:** "Google doesn't comment on unreleased products."

*Story snippet:*

We've heard whispers from several well-placed sources that Google is in the midst of releasing a new line of web-connected refrigerators. Reached for comment, Google spokesperson Aaron Dale stated, "Google doesn't comment on unreleased products."

## On Background

This is how you share information that you don't want attributed to you, but are okay with having the reporter publish. When you see a news article talking about "a source close to the matter," that person was talking *On Background*. This can be used a lot of ways, sometimes quite creatively (and not always in cloak-and-dagger scenarios).

Say you want to explain a complex topic related to your industry (in fact, the reporter asked you to), but you don't want to stress over each word as it leaves your mouth (as you would if you were speaking *On the Record*). Why, this would be a great time to ask whether you can hop *On Background*!

Or: a reporter sends you an email with a bunch of questions, they're clearly in a hurry and you don't want to labor over your written answers (because they might get quoted verbatim). *On Background!*[*]

And, of course, if you want to leak some secret info (there's a whole section on that later, scoundrels). *On Background!*

There's a lot of leeway in how your *On Background* conversations will (or won't) be attributed, and it's something you should

---

[*] Think *Hot Pockets* jingle.

work out with the reporter prior to telling them the goods. For mundane stuff there often isn't attribution at all and the reporter simply asserts what they've been told. If you're dishing secrets reporters will generally push for a more specific attribution; you'll probably lean the other way — though remember the less specific you are, the less credible you become[†].

**REPORTER:** What's this about Google launching a new line of smart refrigerators?

**AARON DALE:** Can we go *On Background*?

**REPORTER:** Sure.

**AARON DALE:** Ok — you can use this, but don't mention I work here, ok? Just say I'm someone with knowledge of the project.

**REPORTER:** Hm. Okay.

**AARON DALE:** So, yes — we're building a fridge. Two of them actually. One of them sends you a push notification when your milk is going bad. It's pretty badass.

**REPORTER:** Awesome. Can I say that? That it's badass?

**AARON DALE:** Ha, sure.

**REPORTER:** Okay, back on the record?

---

† Actually, it's possible for this vagueness to make you sound *more* credible, if your relationship to the company/deal is far removed. "A source" sounds more credible than "a guy who worked briefly with this executive three years ago." Good publications won't go for this kind of shoddiness, but they aren't all good.

**AARON DALE:** Yup. You can say our comment is, "Google doesn't comment on unreleased products."

*Story snippet:*

> We've confirmed with a source with knowledge of the project that Google is indeed working on not one, but two models of smart refrigerator — one of which will actually send your phone a push notification when your perishables are on their last legs. "It's pretty badass," our source tells us. Reached for comment, a Google spokesperson said that "Google doesn't comment on unreleased products."

**Off the Record**

The reporter can't print what you're telling them in any capacity (they can't even print the info and cite an anonymous source). They *can* call someone else and probe about the same topic to try to get *them* to talk about it, which means they can write about it — and maybe that's what you want them to do — but they can't refer to you at all in these subsequent conversations or in print. This isn't as safe as it sounds, because if they wind up calling around and aren't tactful about it, they may inadvertently reveal who put them on the scent.

One way *Off the Record* can be used is to tell reporters about things that you're not allowed to talk about but will help them write a better (or more positive) story.

Say you're talking to a reporter who doesn't sound convinced that your company is as compelling as you're claiming. What they don't know is that you're wrapping up negotiations for a big partnership with Facebook; it definitely can't be in the article, but you want them to have a sense that you're the real deal. So you tell them about it *Off the Record* and — just like that — they're convinced (but: they just ran off to find a source who will confirm this Facebook deal, so maybe you just blew that up, hotshot).

*Off the Record* conversations are also used to tell reporters when they're about to do something dumb.

**REPORTER:** So, I'm writing a story that says Google is following up on its successful line of refrigerators with a line of toaster ovens.

**AARON DALE:** *Off the Record?*

**REPORTER:** You can't say anything, even *On Background?*

**AARON DALE:** No, I'm too close to it.

**REPORTER:** Ok, *Off the Record.*

**AARON DALE:** *Off the Record*, if you write this story, you will look like a jackass.

**REPORTER:** Welp, worth a shot.

Finally: some reporters think *Off the Record* (where they can't write anything you say) means the same thing as *On Background* (where they can).[‡]

‡ This ambiguity has persisted for decades. The late *NYT* columnist William Safire explored these terms in 1989, writing: "Some sources now say 'Off the record' as gossips say 'Don't tell a soul,' meaning 'Pass it on, but don't say I told you.' Others will use off the record to mean 'attribute it to somebody else,' or fuzz it up, which means 'talk to a bunch of other people to cover my footprints.'"[2]

In the late 90's *Slate* conducted a poll of *Washington Post* reporters with depressing (and amusing) results. Among the responses: "Most of the time, when people say off the record, they don't really mean it... I have no idea what 'off the record' means."[3]

# Destiny

*One fateful evening, our beleaguered future-founder Joanna Sagan is banging her forehead against her desk, weary and miserable and ready to scream. All day she's been working on this report; if it isn't on her boss's desk by 5 o'clock she can kiss that bonus goodbye. It was a ridiculous task — insurmountable for most — but she did it, she wrote the whole damn thing without so much as a lunch break, and now, this.*

*How did society manage to develop self-driving cars before it figured out how to make printers that FUCKING PRINTED?*

*Her thoughts are spiraling into darkness, visions of bashing the thing with a baseball bat, then standing in the innards of Mt. Doom, the impotent LCD screen flashing MERCY as she cackles and throws it into the fiery depths — and it hits her:*

*What if we didn't need to use printers any more?*

*What if you hit* Print *and someone showed up at your doorstep a half hour later, document in hand, printed on high-bond paper, magazine-crisp text — a truly superior product, without the suffering?*

*The vision appeared fully-formed: small at first, just her and a few friends driving around the city making the deliveries themselves. Then, later, she's harnessing the power of peer-to-peer — neighbors printing for neighbors. Why should everyone be forced to monitor ink levels when they only need the damn things twice a year for their luddite bosses and out-of-touch music venues?*

*A universal problem with global scale. A multi-billion-dollar market.*

*Joanna knew she'd never be able to work on anything else until this problem — her problem — was solved.*

*That night, GutenberGo was born to change the fate of the printed word.*

# Story

Reporters know founder stories smell distinctly of bullshit. Some are complete fabrications, though, mostly, they're factual-ish moments neatly arranged to spin a yarn*. Despite this, we're happy to include them in our articles. In fact, we *want* to. It's as if there's an unspoken agreement between founders and reporters: make the stories plausible, and we'll act as if they're true.

Because people like stories. Stories make companies worth caring about, and they make articles worth reading. They take the jumble of features you're trying to sell and wrap them up into something greater — or at least more purposeful — than the

---

* To be clear, I'm not saying you should lie, particularly around things that are provable. Live long enough, and someone will check it. Cherry-pick the truth.

sum of its parts.

The founding story is just the first act. You need to keep building on it — through product, through press — to further reinforce this idea that you're working on something bigger than yourself. You're on a mission, damn it, and if you aren't telling us what that is, no one is going to do it for you.

All too often, companies will pitch themselves as improvements on some big-name incumbent without explaining *why* any of it matters. Why do people need this not-so-different app from what they're already using? Why are you the people to bring it to them? Why can't the big guys just copy you?

The answers to these questions are not objective facts. They're narrative.

You know people are yearning for this because you've experienced the problem yourself. You've got a team with tech chops honed by top schools and elite ex-employers. You're passionate about this in a way no one else is. And sure, the big guys could crush you, but it'd be just another feature for them out of dozens. It's not their core competency, they are't *focused* the way you are, working on fixing this thing since day one.

People may roll their eyes, but they can't prove you wrong. Which means you've got a chance.

# Mission vs Experience

Real quick: in Joanna's founding story, we learned she was inspired during a bout of extreme frustration that led her to GutenberGo. This is fine — Dropbox founder Drew Houston was driven to build the multi-billion dollar file sync service because he got sick of using thumb drives — but as stories go, this first act is just average.

What if, instead of simply having a bad night at the office, Joanna ran a major printer-repair operation where she regularly bore witness to the pain and suffering these archaic machines inflict on their users (who subsequently took it out on her)?

Now we've got someone with intimate knowledge of *why* these things are so unreliable, has seen the suffering they cause, and knows how to build a heavy-duty alternative she can distribute to anyone who wants to serve as a GutenberGo partner. She's on a mission to make the world a better place, and she's better equipped than anyone to make it happen.

That story's a lot more compelling. It's not the one we're going to run with — and I'm not saying you need one this strong — but I wanted to demonstrate that there's a difference between a company sparked by moments of frustration and missions propelled by domain knowledge.

Plenty of great companies get founded by people new to their respective industries, but there are scenarios where having a lot of experience helps.

- If you're dabbling in a gray area of the law, a position that exposed you to how the industry is regulated would be a boon.

- If you're reliant on deals with major content owners, you'll look a lot more credible if you've worked with them before.

- Are you offering doctors a way to send their patients advice? It'd be better if you were a doctor, not just a patient.

Exceptions abound to these. But if there's stuff in your background that can bolster your story, use it.

# The Big Question

To recap: your story is vitally important and will dictate how you talk about yourself, write about yourself, and steer your company. But when it comes to writing your initial pitch email, don't spend more than a couple of sentences on it — something to establish your credibility and focus, and move on. You'll expand on the other details later.

Which brings us to the second key component you need to include in your pitch. It's best expressed as a question, and it's one I ask in nearly every interview — but the good pitches beat me to the punch.

*How is what you're doing better or different than whatever is already out there?*

Your answer to this may sound similar to the story you just established; the two ideas dovetail nicely. But in this case I'm looking for something more concrete: how your story is expressed in the product.

When I first lay eyes on your pitch I'm trying to figure out three things:

*Who are you? What does this thing do? Why should I care?*

Your story should cover the Who and Why just fine, but the middle one — what you actually *do* — is less obvious than you'd think. Too often, startups get caught up with their hand-wavy missions without making it clear how their dreams have manifested themselves in anything remotely tangible.

And so, the question:

*How is what you're doing better or different than whatever is already out there?*

Which is really two questions.

Let's start with the second: *"whatever is already out there"*.

This is me trying to get a sense of context. If you compare yourself to an extant product — even if it's just a vague reference — you're giving me an easy jumping-off point that doesn't require me to mentally reinvent the wheel. Depending on who your competitors are you don't even necessarily have to name them. For example:

> "There are several solutions for letting consumers outsource their printing needs, but they all suffer from the same problem: they require too much work. We're fixing this by…"

> "FedEx/Kinko's will print your documents and deliv-

er, but we've turned their model on its head by by-passing the inefficiencies of retail outlets..."

A picture is starting to form. Now instead of contrasting you with everything under the sun, I'm zooming in, coming that much closer to figuring out what you do, who else is doing it, and, ultimately, whether it's interesting. And since you're pre-emptively framing all of this in such a way that it *is* interesting, your odds are that much better.

Which brings us to the other part of the Big Question: *"How is what you're doing better or different?"* — the nuts and bolts. This is where you'll describe the big differences between yourself and the rest of the competition. You'll want to keep it short, without a lot of explanation; bullets work nicely. For example, Gutenber-Go might say:

- Our mission is to get you your document where you need it, when you need it. Delivery is typically within 30 minutes (guaranteed under an hour), and we let you see exactly where your document is on a map, in real-time. (FedEx gives vague five-hour windows and costs 40% more).
- We're focused on quality. We use a custom high-bond paper (32 lb) because we think it makes all the differ-ence. No one else does this.
- The other guys make you jump through a hundred hoops — business accounts, color settings, the works — to print anything. We take three taps.

Context is king, so strong pitches often include data and ref-erences to the competition in the bullets themselves. Don't just say you have a feature or boast about a stat — explain why it's impressive by pitting it against what's out there, otherwise you're leaving it up to the reporter's prior knowledge.

A warning: it's easy to make these bullets sound like a hodge-podge, which feels confusing and desperate. Most of the bullets should double as supporting evidence for your story. It's okay if you have a point or two discussing features that are genuinely cool and only tangental to your mission, but make sure it's icing on the cake.

# Clichés

### *We're X for Y*

The last section talked about establishing context, and I want
to address the most common way that startups do this: by em-
ploying the startup trope that reads, "We're X for Y" — as in,
"Airbnb for dogs" and "OkCupid for platonic friendships".

If you're ever tasked with writing a joke about startups, this
is probably where you start. These snippet-sized pitches often
sound like oblivious self-parody, and even the sensical ones are
geek-babble — it's no wonder some reporters actively crusade
against them.

I get it. I'm sick of them too. But if I'm an entrepreneur and
you stick me in front of an exhausted reporter who will give me

ten seconds before they zone out, there's a good chance I'm doing exactly the same thing. *Maybe* I can frame my product in such a way that I'm talking about the problem it's solving without referencing anyone else, but not everything is that straightforward, and sometimes there are genuinely good comparisons that make things easier on everyone.

Anyway, for the purposes of your pitch, here's my advice: don't use the "X for Y" format verbatim. Spread it out a bit so that you provide a basic description, then compare it to something else, add a little context, maybe poke fun of the fact that you're relying on this trope — but if you can't figure out a way to describe yourself in less than a sentence, you could do a lot worse.

And if your pitch doesn't lend itself well to one of these comparisons, don't force one. It'll just confuse people.

### Z That Doesn't Suck

Startups take this approach when they're building a product that's functionally better (they think) than an incumbent. So, *Office that doesn't suck. Chat that doesn't suck.*

I can't pin down exactly why I don't like this. Maybe it's because it's overused (but: I just defended another cliché). Maybe it's because it's often hostile toward another company (but: some deserve it). Mostly, I think it's because it's indicative of a company that doesn't have a bright future.

When you say you're *"Office that doesn't suck"*, then you're really saying that you're competing on features. Maybe you've got a more intuitive interface, more options, faster sync — whatever it is, it's an improvement on something that's already out there.

There's a business maxim that goes like this: in order to get people to switch from whatever they're using, you need to be eight (or ten, whatever) times better. Not 80% better. 800%.

What it's getting at is that people are heavily influenced by inertia. They're going to stick with whatever they're using because they've already invested in it (both in terms of money and hours) and they know where the buttons are. Sure, there are things they'd love to change, but for all they know you have your own set of pitfalls (and since you're a startup, you do).

So unless you're a huge leap forward, you're probably taking the wrong angle. Better to craft a story that explains not just that you're better and different, but how you cater to particular groups of people and use-cases. It's a subtle difference, but it matters.

# The Competition

A bit earlier I said you should include context by referencing your competition. Turns out it's complicated, and I want to tackle the two most common issues.

### The Ostrich

The Ostrich founder does not necessarily deny that competition exists, but would prefer it didn't and will swerve around any question implying that it does. They are averse to comparing themselves to any existing product — and their pitch is devoid of any such comparison — forcing the reporter to tease out every little detail (more likely, I'll get annoyed and give up).

Let me be clear: acting as if you have no competition is a bad strategy. Yes, there is a remote possibility that you are frolicking in a greenfield opportunity with no one else in sight, but you're probably bullshitting me, and I don't like that (also: there are founders who seem to genuinely not know about their competitors, which is depressing and a good way to make me reconsider covering you at all).

You could argue that you don't want to discuss your competition by name because you don't want to give them free press — the people reading the story may not have heard of them — and that's reasonable enough. But you can still talk about them in vague terms ("There are some others in this space who are doing…"), which is loads better than, "We don't think anyone is approaching this problem the same way we are." Um, sure.

And know this: even if you make a point not to mention your competitors, good reporters will name them anyway in their articles. In fact, it'd make me try that much harder.

### The Bully

Making aggressive attacks on another company is risky and can backfire badly. It's hard to get the tone right, and unless there's consensus that the product you're attacking is, indeed, pretty sucky, it may sound obnoxious (if you do make a pointed attack, make sure you're punching upward).

The trick is to tear them down politely. You don't need to use mean words to make them look bad, when you can just as easily cherry-pick facts and anecdotes that do the same. Let's look at an example.

Say GutenberGo has a competitor called PrintDriverz that promises to deliver your printed documents within 24 hours. Aside from being slower than GutenberGo, Joanna thinks PrintDriverz's software is pretty shoddy. As a founder she might be inclined to write or say something like this to a reporter:

"The only other company with a mobile app is Print-Driverz, and they're slow as molasses (we take an hour, they take 24). Very few people have the foresight to print anything a full day in advance, and their software is notoriously buggy."

(I'm assuming you intuitively know it's a bad idea to throw serious vitriol around, so I made this 'attack' tame to illustrate that it can still make you sound like a jerk).

Now let's say I'm the reporter and include the quote in my post. That might look something like:

Founder Joanna Sagan says that PrintDriverz is "notoriously buggy" and "slow as molasses", and that GutenberGo aims to address both these issues by building out an extended infrastructure...

This has two problems. First, Joanna sounds like a bully. If that's your thing, fine, but it's a bad fit for most founders. Second and more important is that the attack is flimsy. It may resonate with people who already aren't satisfied with PrintDriverz, but it'd be more credible if the conclusion around PrintDriverz's issues came from the reporter's words, not hers. A founder who doesn't like their competition? *You don't say.*

Let's try a different tack. Instead of pointing out PrintDriverz's flaws herself, Joanna will merely present the evidence and let the reporter reach their own conclusions (which are, hopefully, exactly what she steers them to be).

So maybe she says this:

"We've focused on making GutenberGo the fastest solution out there. Our guaranteed delivery time is under an hour, and 80% are delivered in under 30 minutes; PrintDriverz's window is a full 24 hours. We've also engineered our apps to be native to their platforms, while PrintDriverz relies on Adobe AIR."

And here's where she can do some more advanced stuff, like talking *On Background*. Say we're in the same interview — the nicer one — and Joanna really wants to hammer home that PrintDriverz is a turd. So she says to me, "Is it alright if I go *On Background* for a moment?"

And me, the reporter, I'll probably say, "Okay."

And then she says, "I don't want to bash them on the record, but PrintDriverz's app is incredibly buggy. We've heard this feedback from our customers, and you can verify it by looking at complaints on Twitter." *(After the interview, she'll follow up with an email linking to a demonstrative Twitter search query\*).*

And I'll say, "Gotcha. Ok, back on the record?"

And she'll say, "Yup." And off we'll go.

And now my post looks like this:

> "GutenberGo promises to deliver your documents in an hour or less, with 80% of them delivered in under 30 minutes, compared to PrintDriverz's relatively sluggish 24-hour window. Founder Joanna Sagan adds that the company has focused on app quality — in contrast, PrintDriverz has drawn sustained criticism from its users *(and here, I link to some of those nasty tweets)."*

Broadly speaking, your competition is a topic you'll get asked about a million different ways, and sometimes the one-liners

---

\* This wouldn't work if there were only one or two complaints, of course. Three, well...

you memorize won't cut it. In these cases, consider talking about them as if you believe the following:

> It's not that they're a *bad company* — they've got good people working hard, and, as it happens, you're friendly with the founders — but they just don't have the technical background and focus you'll use to bring this thing to the mass market (or, if they already own the mass market, they've let the inertia of big business keep them from staying innovative).

I know that sounded sardonic, by the way, but I actually think it's a decent mindset to have. As humans we have a tendency to vilify the opposition, but your competitors are probably good people working hard to make something they'd like to be proud of, so don't be too quick to shit all over them.

# Choose a Mark

You can't lock in your pitch until you figure out who you'll be pitching, so start thinking about it early. Tone and topics vary widely between publications, and — sometimes more so — between writers at the same publication. Be warned: these sites thrive in part because many readers have relabeled procrastination as "research".

Some questions to ask as you conduct said "research":

- Who writes stories that you actually finish reading (or, let's be realistic: make it half-way through?)
- Who covers companies similar to yours?
- Does it seem like they have a high bar for 'interestingness', or do they get worked up over incremental fea-

tures?

- Do they cover new startups often? (Senior writers trend toward the big companies and startups that already have lots of buzz — they're generally more important, and they're often more fun to write.)

A single post does not indicate a trend; just because your favorite writer fawned over a direct competitor doesn't mean they'll do the same for you. Reporters sometimes dip out of their normal range for reasons that have nothing to do with the story itself (they've got a date in an hour and need to get one more post up, and, hey, this'll do), so don't bet the farm on a possible fluke.

Eventually you'll have a list of writers you like. Figure out which of them are most established and put them in the 'reach' column. Popular writers get the most pitches, which means they're more reliant on intros from trusted 'legit' people like VCs, veteran entrepreneurs, and good PR reps — and your cold-pitch email has a greater chance of going unseen.

Thankfully there are always great writers who haven't built huge followings yet. Keep an eye out for unusually good new-comers, who are likely receiving fewer pitches and are less jaded (and more easily excitable) than their senior colleagues. Writers covering less consumer-focused beats are often under-appreciated as well. And of course, if the site you're pitching has a reporter dedicated to covering the sort of thing you're building, you'll probably want to go with them.

Avoid the jerks if you can, but I'd take a snarky post that gets heaps of comments over a ghost-town press release rehash any day.

# Embargoes

If I wind up in Hell then I have a pretty good sense of what to expect. It looks like my work desk: computer in front of me, phone next to it. Every five minutes or so (at semi-random intervals, so I don't know exactly when it's coming) the phone rings. It keeps ringing louder til I pick it up, so eventually I do, and at the other end a voice says, "I've got something I'd love to tell you about... do you accept the embargo?"

Madness — rats-knawing-on-my-face madness — quickly ensues.

Not that this matters much to you (maybe you think I'm getting off easy), but I wanted to give a sense of how much reporters loathe embargoes. They're the worst. Unfortunately they're a fundamental part of the press cycle, so take my hand and let's head down this path of drudgery before I find an excuse not to. Be warned: a misstep here can destroy your credibility with every publication and reporter you're liable to encounter. No pressure.

Let's start with some definitions.

An *Exclusive* is a promise to a reporter that you will give them, and them alone, the news you are pitching.

You can think of an *Embargo* as the opposite of an *Exclusive* — because it often involves briefing multiple reporters rather than just one — but the distinction is more nuanced than that. The real definition of *Embargo* is this: an agreement made by a reporter to withhold the story they are writing until a specified date and time. I'll make this clearer shortly, but let's tackle the easy stuff first.

If you're running your own PR, then the *Exclusive* is your friend. It only involves one reporter so the logistics are relatively straightforward, and it increases the likelihood that the reporter will be interested in what you're peddling because they know they'll be the only ones with the goods. In fact, this ownership over the story boosts the odds that they'll write something *positive*; it's not guaranteed, but it helps (this is one reason many 'negative' stories, like disclosures of data breaches, are often given to reporters as exclusives — they're less negative than they would be otherwise).

There is no trickery involved in offering a reporter an exclusive. You email a short version of the pitch, the word "Exclusive" featured in the subject line and perhaps bolded in the message itself. You don't want to intimidate them with a wall of text — you'll send a second, more robust email if they're interested — but you'll include more detail in this first message than you would if this were an embargoed story.

Cross your fingers and wait awhile (at least a day before you try again). With any luck they'll give the thumbs up, you'll follow with a more comprehensive rundown, set up an interview, and — bam — press.

Actually, there is a little opportunity for trickery. Should this reporter decide not to cover the story you've offered them exclusively, you can pitch someone else as if they are your one-and-only. Make sure you've gotten the first guy to say "No" before you try, though — you'll be dealing with a couple of incensed reporters if they both write stories under the impression they're the only ones doing it.

Still with me? Feel free to get a drink (I just did). Now for the hard part.

Let's say GutenberGo is looking to get the broadest coverage it can, across multiple publications. News sites have no interest in being 'late' to a story — if one of their competitors has already written it and it isn't huge, everyone-needs-to-cover-this news, then it's effectively dead to them. This obviously makes appearing on more than one site a challenge. The compromise: sometimes they're willing to agree to be tied for first.

Enter the embargo.

In theory an embargo can be a mutually beneficial arrangement. Reporters are given ample time to process and research the story, conducting interviews with GutenberGo executives (perhaps even some industry experts), and to write something with more substance than a knee-jerk under the gun of 'breaking news'. GutenberGo benefits from having more accurate, well-written pieces about them — and what they really care about: coverage from more outlets.

During an initial 'feeler' round of emails, Joanna will dangle a hint of the news in front of several reporters, promising more details if the reporter is willing to agree to an embargo of, let's say, 9 AM PT, Tuesday the 10th. This initial email can't have *too* much detail — just a whiff that there's something solid behind it — because otherwise a reporter could potentially run off and write the story immediately.

Reporters that decline get scratched off the list ("he's a boring writer, anyway…"), those that accept are sent a full pitch, screenshots, an offer for an interview, and whatever else they might need before the newly-set deadline. If all goes to plan, everyone's post gets published at 9AM PT Tuesday the 10th on the dot (reporters can 'time-stamp' pieces to go live at a specified time, without having to hover over their computers).

But reporters almost never like embargoes (who wants to be tied for first?). In some cases they don't have much choice — they're going to be writing about Google and Facebook's new stuff anyway, and they can either accept their embargoes and be tied for first or say "no thanks" and come in 40th. But Gutenber-Go doesn't have that kind of leverage: a reporter can easily turn down their pitch and nobody is going to bat an eye.

So, GutenberGo's success at getting reporters to agree to an embargo will be determined by the following factors:

- How cool the product is. If it's something that's already topping the charts, or is brand-new and is really going to blow everyone's socks off, they have a shot. And if it's that cool, they probably also have…

- Big-name funding. Having a list of impressive investors is validation that makes it much easier to get reporters to accept an embargo. Alas, GutenberGo doesn't have either of these yet, but they still have a shot if they get lucky on a…

- Slow news day. Reporters are expected to write a certain number of posts per day (even if there's no set number, everyone has a sense of what your output looks like). Sometimes it's easier to accept a Friday-morning embargo than to spend the whole day scouring for something vaguely newsworthy (not much happens most Fridays). Of course, this 'trick' isn't a secret — other people are trying to send embargoed pitches too.

Put another way, asking for an embargo dilutes the appeal of your pitch. If you're a startup that's never received coverage and has no big-name investors then you'll have much better luck offering your story exclusively to one reporter. Maybe a great PR person can eek out an embargoed launch for you (in which case they'll be leaning heavily on their relationships), but it's tough.

But let's say Joanna and team are confident that what they're doing is going to be a press darling. It makes for a great demo, has early buzz, and Joanna is on friendly terms with a few reporters. It'll be a pain, she knows, but it's time to go big. It's time to coordinate that embargo.

So she does what we just discussed — the feeler emails, the full-pitch follow-up and pre-briefs with reporters who accepted — and things have gone well so far, with four reporters lined up to write the story. Just another twelve hours til GutenberGo makes its big debut. Which is when everything goes to shit.

At 9:05 PM, the night before the big launch, Joanna's phone vibrates. "Congrats!" a friend says. She pauses for a moment, confused. Another vibration, this time a push notification from Twitter. There's a link on it. To the article that just announced her launch. And holy shit is this bad, because the servers aren't even up, the app isn't live in the App Store, and the panic is rising in her chest as —

Her phone is ringing again. She recognizes the number but can't remember why.

It's me, one of the reporters she briefed on GutenberGo this afternoon. That isn't my story that just went live, because I don't break embargoes. I'm upset.

"WHAT THE FUCK?!" I say. I'm unpleasant when I'm upset.

"I'm sorry, I'm sorry, I don't know what happened."

"Why the hell did you brief *Mashable*? Who briefs *Mashable*?! What the hell are you thinking?!"

"I'm sorry, I'm so, so sorry. He agreed to the embargo, I don't know what happened, I'm—"

"Yeah, well, thanks for wasting my time," I say as I violently push the 'End Call' button on my phone, hoping she can hear the tap.

Meanwhile Joanna is a mess, has no idea what went wrong. She opens her inbox and sees an email from pissed-off-reporter #2 with language that makes me sound like Mr. Rogers, then a barrage of tweets from frustrated strangers trying to get to the app that isn't there. Finally she gets it together enough to call the guy who posted his article twelve hours early.

"Hey, sorry, was just writing you an email," he says. He sounds sorry, but not that sorry.

"What happened?"

"I screwed up, had another startup post that was supposed to go live tonight at 9PM, and I got that one mixed up with yours. I can take it down if you'd like." It sounds like he's eating something mushy.

And *The Ghost of Embargoes Past* whispers, "I think we've watched enough of this train wreck," and in the distance we see that it's going to keep burning for quite a while.

........................................

So that's the rub with embargoes. Conceptually they aren't difficult. Logistically they're a pain, but no more so than other logistical juggling — except that when you drop a ball it explodes, and all the other balls explode too, and sometimes they explode on their own accord, and maybe they don't kill you but your eyebrows are gone and who knows if they'll ever grow back.

Despite this, embargoes are an everyday occurrence, and good PR people can usually manage them without incident. It's possible to conduct one yourself, too, but you need to be careful and keep in mind that there is no way to eliminate the possibility of disaster.

With that said, here are some tips for running an embargo:

- Always include the date, time, and time zone whenever you are discussing the embargo. Make it bold. Is Daylight Savings this week? Point it out in big red text.

- Seriously: Time Zones. Reporters have a depressingly hard time with them and they are responsible for many broken embargoes. For US-based tech reporters the standard tends to be PT, but include their local timezone as well.

- You need each reporter to explicitly agree to your embargo. Under no circumstances should you send an email that says "This is embargoed" and imply that they are automatically held to it (it's a common pet peeve).

- Keep track of what's happening. Obviously you have

some logistical overhead when dealing with exclusives, too, but each additional reporter adds a bunch of ways to screw up. Use a spreadsheet.

- If someone *does* screw up the embargo, immediately call and email every other reporter and tell them they are free to publish (while apologizing profusely). Doesn't matter if your app isn't on the App Store yet. Cat's out of the bag. You can try to get the person who originally broke it to take their story down (it might make the other reporters feel a little better), but in this era of caches and rehashes it's always seemed a fool's errand.

- Some sites break embargoes as a rule. Don't ask them to agree to an embargo.

- The more reporters you brief, the more likely it is that everything will fall apart. This isn't just a matter of flipping the coin one too many times — as you cast a wider net, you're liable to start talking to reporters who are either less experienced with embargoes or have less to lose if they botch one. Reporters for lower-tier sites have actually broken embargoes on purpose, because the penalties (probably limited to some snarky tweets from other reporters on Twitter) are overshadowed by the page views and attention they generate.

- If I were doing an embargo, I'd keep it to two or three reporters. Maybe four, if they're all super-reputable.

- Some reporters will ask how many people you're briefing. If you tell me "three", that makes me like you a lot more than "a handful". But if the number is above three, then "a handful" is probably the better choice.

- I want to flag one perk of a successful embargoed launch

(there are some, despite what I may have suggested): when a story appears in multiple publications, it makes it seem inherently more important (in other words, it isn't just about getting more eyeballs).

Whew. We're almost done.

Earlier I said that an *Embargo* (where you brief multiple reporters) is not quite the opposite of an *Exclusive* (where you brief one). That's because it's possible to put an embargo on an exclusive story — say, by asking your lone reporter to wait to publish their piece until your servers are ready on 9AM PT Thursday. This is often fine with the reporter, but you want to be careful about how you use the word "embargo" in this context. Instead, consider a friendlier-sounding euphemism, like asking them to "hold their story" til then. Embargo isn't a word that makes reporters feel warm fuzzies, and they may assume you're talking about one of the briefing-multiple-reporters doozies, so if you use it, make it glaringly clear that it's theirs exclusively.

Finally — this is important — if you give a reporter an exclusive you can offer them flexibility on when they publish their post. So, instead of asking them to "hold til 9AM PT Thursday", you could say "a post any time Thursday or Friday morning would be great" or even, "Whenever is good for you."

As a reporter I *love* this — I'm never sure when there's going to be breaking news, so a good story that I can get to when it's convenient for me is a godsend and may improve your chances of coverage. That said, affording too much flexibility can stick you in an endless purgatory of the reporter intending to "get to it soon, hopefully tomorrow!" (it won't be tomorrow).

Final, final thing. My vision of hell earlier, with the phone calls? Scratch that. It's writing 2,500 words on embargoes.

# The News Cycle

The news waxes and wanes in a predictable pattern.

Monday is a crushing onslaught that seems all-the-more forceful after the relative tranquility of the weekend. The torrent continues Tuesday; Wednesday holds strong but shows signs of abating, and the cycle decelerates through Thursday and Friday.

Each day is a cycle unto itself: mornings are heaviest, afternoons are lighter, and some PR people will beg for their stories to be published by 2PM PT, before the East Coasters leave work.

Friday mornings may see some action, but the waves ebb by noon as we all check out for the weekend. Watch closely come Friday evening, and you'll see some corporate goons shoveling their trash into the receding tide just as the sun sets.

Too flowery? Here's what you should know as a startup:

- Be wary of launching on Monday or Tuesday mornings. They're noisy.

- Keep an eye out for press events. Some, like news at big conferences, are easily avoided; others are last-minute and bad luck.

- Big news tends to break in the morning, so consider late morning and early afternoon, Pacific Time. Not only will you have less competition, but reporters will be finishing their first stories of the day and looking for their next fix.

- A story published in the afternoon or evening is miles better than no story.

- There is a time-honored tradition of "dumping" bad news on Friday afternoons in the hope that it fizzles over the weekend. Reporters know about this, so don't make it blatant, but there's a reason it's a tradition.

The above is relevant to when you send your pitch emails, too. Don't send these during peak hours or major press events.

Do what you can, but it's impossible to completely avoid running up against big news — the Googles of the world release stuff throughout the week with no warning. There is one exception to this: big companies rarely have news during holiday breaks (particularly the two-week stretch around Christmas when everyone seems to be on vacation). Many reporters are still working during these breaks and readers are yearning for anything to distract them from their in-laws. If I'm a startup, I'm pitching a story every holiday season (which doesn't mean it has to be *about* the holidays, mind you).

# First!

As I've mentioned a few times, reporters care a lot about being the first to write a given story. We also care about getting attribution in other articles — "...*The news was first reported by The New York Times*," for example.

And when I say we care, I mean we behave like petulant children when we don't get our dues. Follow a reporter long enough on Twitter and the probability of seeing them complain about some inside baseball journalistic injustice approaches one.

Part of this has to do with credibility. If I'm first to a big story that everyone else cites, then I get points from the sort of people who pay attention to bylines (smart readers do — I'd take my half dozen favorite writers over any of the big sites, any day).

So the incredulity makes sense in the case of substantial news, call them doubles and up. But most news breaks are down-by-seven singles. They move the story forward, but no one remembers them an hour later. Except, of course, for the reporters who wrote them.

Because we want to win. Reporters can talk about serving the public all we want, but we're fueled by a primal thirst for victory. Every scoop I get is one my competitors didn't, and even the small ones feel good, if only because I know my rivals are ticked off at not breaking it themselves (more likely, their tender egos will murmur, "how is this even *news?*").

This is one reason shoddy (and vile) stories get published: the desire to 'win' overtakes credibility and morality. And it's why you should never break your own news.

Like I said, inside baseball. Just help us get to first.

# GutenberGo's Launch

What follows is a rough timeline for GutenberGo's launch process, assuming they opt for giving a single reporter an exclusive. Drumroll please.

**T-MINUS:** *Months before launch*

Joanna and her cofounders begin talking about press strategy. They set a target launch window, hash out what they're planning to have finished by then, and are getting a feel for how to articulate their company's story. They read too much tech news — "research" — but at least keep an out for bylines whose stories are well-done and aren't focused exclusively on big companies like Google and Facebook.

Joanna makes time to attend various startup cocktail hours, again with a sense of purpose. She networks and keeps an eye out for reporters. If she spots one she tries to strike up an easy conversation but knows better than to give her whole spiel; no way they'll remember it. Better to keep it short and friendly (there's still a ways to go before launch) and ask if there's a good way to reach the reporter in the future.

**T-MINUS:** *Three weeks before launch*

Joanna sends her favorite reporter a tip on something unrelated to her company. We'll discuss this later.

**T-MINUS:** *Five days before launch*

Joanna sends an email to her favorite reporter. Let's suspend reality and make that me. Her email — which contains a taste of the news designed to entice but not overwhelm — reads:

### *(Exclusive)* Chuck your printer! GutenberGo prints your documents on demand

Hey Jason,

Wanted to check in about GutenberGo, my company that I mentioned when we met at StartupSiege a couple of months ago. We're launching next week and I'd love for you to be the one to cover it exclusively.

Here's the gist:

- GutenberGo is the answer to printing headaches. Tap a button on your phone, tablet, or computer and someone is at your doorstep, document in hand, in under an hour.
- We're cheap — orders are $7 for up to 10 pages.
- Of our beta users, 90% who placed one order returned for another.

We're still prepping the servers, so it'd be great if you could hold the story until **Wednesday, March 11 at 9AM PT** — please let me know if you're okay with that and I'll send over screenshots and more details. I'd also love to swing by your offices or jump on a call to walk you through the details.

Joanna
CEO, GutenberGo
925 555 1918

**T-MINUS:** *Three days before launch*
Joanna hasn't heard back from me (I'm rotten at email). She writes:

> Hey Jason, just wanted to ping you in case this fell through the cracks earlier. Would love to tell you about GutenberGo! (Also: I totally understand if it isn't up your alley, but please let me know either way so I can take the exclusive elsewhere).
> Thanks!
>
> Joanna
> CEO, GutenberGo
> 925 555 1918

Fifteen minutes later, a bit spooked by the friendly threat to give the story to someone else, I respond:

> Hey Joanna, sorry, been swamped. Yep, I agree to the embargo, sounds interesting. Can we set up a call for Tuesday, say 1PM PT?

Joanna immediately shoots back:

> 1PM PT sounds great! I'm at 925 555 1918. I'll follow
> up before the call with screenshots and more details.
> Looking forward to telling you about GutenberGo!

**T-MINUS:** *Two days before launch (the night before the interview)*

Joanna sends the following email, which is more fleshed out than the last one (in fact, it has enough information that I could potentially write my article without bothering to pick up the phone). It's okay for this email to include some redundancy, no reason to make me go digging through the thread:

> Hey Jason! Looking forward to our call tomorrow
> at 1PM PT. Here are the main things to know about
> GutenberGo (I've included several screenshots as attachments, and you can find a video showing off the
> product right *here*).
>
> For decades, printers have been a thorn in our sides,
> plagued by mechanical errors and buggy software that
> never seems to improve. Printer usage has decreased
> — we can pull up most documents on our smartphones — but few of us have been able to escape them
> entirely (there's always that one venue that requires
> paper tickets). That's where GutenberGo comes in.
>
> • GutenberGo will print any document and have it
> ready for you, at your doorstep, in one hour or less,
> guaranteed. 70% of documents are delivered in 30
> minutes or less.
>
> • Documents look incredible: they're printed on high-

bond, 32 lb paper, which is far better than what most people have in their homes or offices (the words 'tasteful thickness' come to mind...)

- We charge $7 per delivery for any combination of documents up to ten pages; $12 for up to fifty pages.

- Our beta testers are giving rave reviews; 90% who completed one order returned for another, and we've grown 200% week-over-week by word of mouth alone.

- Our CTO Huxley Dawkins was previously tech lead at Xerox's Printer Quality Group.

- The company got its start when founder Joanna Sagan realized that the real solution to *'Paper Jam'* isn't another bout of frustrating gear-cranking and hair-pulling — it's getting rid of the printer entirely.

Here are some screenshots, please let me know if you have any questions, and looking forward to talking tomorrow!

**T-MINUS:** *One day before launch (interview day)*

We do the interview, where Joanna goes into more detail and fleshes out the story (more on that later). I call 10 minutes late but Joanna is gracious about it. She follows up with an email thanking me for my time, and clarifies one point we discussed that she had to double-check a stat on.

### Launch Day

The post goes live as scheduled. Joanna checks it for errors, notices a couple, and sends me an email thanking me for the great post — and if I could please fix the broken link to GutenberGo's site, that'd be great. She shares the story broadly on social media and encourages friends to do the same, visions of an IPO glimmering in the distance.

A few notes about this exchange:

- Joanna always includes her phone number.
- She includes the calendar date and time zone when discussing the embargo.
- She replies quickly to my responses and is agreeable to my suggestions (i.e. the time of the phone call). This tip is obnoxious because it implies that the reporter should be treated as a superior, but each additional exchange it takes to lock in that interview is another chance for the thread to get lost among countless others. Incidentally, it isn't a bad idea to make reporters feel superior, if you can be subtle about it.
- In the second, "ping" email, Joanna hints that she'll take the story elsewhere if she doesn't hear back soon. Getting the tone right on this is key — if you sound entitled or arrogant then the reporter may just whisper, "piss off" and move on to the next thing, but a gentle nudge can do wonders.
- That follow-up email after the interview can be important, depending on how things went. If there's something you're worried you didn't make clear, you can try again — but don't try to backpedal. These clarifications often aren't necessary (a quick "thanks for you time, please let me know if you have any questions!" is fine).

So there you have it: an unusually straightforward press pitch. It's unlikely yours will go this smoothly, but hopefully the other sections will leave you better equipped to adapt as needed.

# Lift-Off

The post is live! Take a breath and look around you, remember the expressions on everyone's faces, the disgusting mess of take-out on your desks, the rush of nerves and fate you feel right now. It'll be gone soon just like every other moment, but you'll be able to look back on this mental snapshot one day and even if it isn't with a smile, it'll still be, I dunno, something.

Now: get busy. Tweet and Facebook and whatever other social app is the hip thing, get your friends to do it, too. Submit to the hot news sites, get those friends (who are getting annoyed by now) to upvote you. Send it to your family, bask in the validation as they see your hermitude has led to something concrete and that thousands are taking notice.

Keep the servers up. Watch for questions about your product, both on the original blog post and elsewhere. Respond to the reasonable, stay cordial in the face of naysayers, and feel free to ignore the bottom of the barrel.

Now prepare to return to obscurity in... let's call it two days. You may have some email threads that go on longer than that, maybe some will turn into something, but this fervent buzz is fleeting and will leave you feeling empty. At least you'll have that moment.

# The Drumbeat

Whew, launch over. I'd say give yourself a rest, but you're a startup so you'd just scoff and sigh.

If you didn't have any luck getting someone to write about your launch it's not the end of the world. You'll still post about it on your blog (more on that later), and there are plenty of other ways to make your press debut. Remember, this is all a crapshoot. Maybe you caught everyone on a bad day.

If you did get press, enjoy it, but don't get comfortable.

Because — woosh, just like that — we're already onto the next story, and with some luck you'll feel that buzz again before too long. And here's some good news, for once: the next one may be easier than the last go-around. Remember, the press has a tendency to reward the already-successful; the more coverage you get, the easier it is to get more.

Now the reporter knows your name and company — it's the start of a relationship, really — so your pitch will jump out from their inbox. They know their way around your product, so it'll be easier for them to wrap their heads around whatever you're doing next. And since most stories include big chunks of context — information that isn't new, often rehashes of stories they wrote before — it's that much quicker to crank something out about GutenberGo 2.0.

This is where the idea of a 'drumbeat' of press comes in. By choosing one reporter (or, if you're gifted at politics, a small handful of them) to regularly send your news to, you'll become a story arc in their minds. You won't have to try so hard to explain why you're relevant. With a lot of luck and a compelling product, you can land in a top blog every month or two.

But there's a reason few companies manage this: startups have a hard time coming up with good pitches that often, and those that try usually do themselves more harm than good.

### The Pitfall: Over-Pitching

Now that you've had a favorable interaction with a reporter, it may be tempting to think that their bar has lowered a bit. I just said that it's easier to get press with them, didn't I?

Allow me to refine that. Your followup press may be easier because reporters are *familiar* with you, not because they're so eager to hear about the next little thing you do. A lot of startups screw up this distinction, sending reporters updates every two weeks on incremental features that make for weak stories. Maybe they'll tell you to wait and "ping them when you have something with more meat", but this pestering will lose you any goodwill you've garnered real quick.

The truth is, the bar *is* a little bit lower for companies that have been previously written about. But you're not in a good position to judge bar height — it's like arguing with an umpire about his strike zone — so you should aim for the strongest story you can.

Your goal is for reporters to not only view your name with familiarity, but for it to shoot a little jolt of happiness into their brains. Not because they're particularly fond of your product (though it helps), but because they have come to associate your name with a solid story — which means they have something to write about, which means they can leave work earlier.

So what makes for a solid story? Why, I just happen to have a few ideas.

# Common Stories

Below, a stroll through the most common pitch themes, including snippets of a sample pitch and its resultant article.

**Funding**

The easiest of non-launch pitches. Funding stories are inherently newsy, writers can churn them out in their sleep, and they don't take much finessing on your part so long as the number next to the dollar sign is big. But they're boring — most funding posts could be distilled to a single sentence with remarkable fidelity — so make sure to include some color (optimistic insight into how you're planning to use the money is a good bet).

*Pitch:*

**GutenberGo just raised $1.4 million from Platitude VC, angels — want the scoop?**

Hey Jason,

First, thank you again for covering our launch last month, things have been going great. We've just closed a $1.4 million round with some great folks — Platitude VC included — and wanted you to have the exclusive if you're interested. Below are some of the details, let me know if you have any questions and want to hop on a quick call!

- Our investors include Platitude VC (Nelson Kovacs is joining our board), and angels including Shari Sarafa, Carolyn Lee Swain, Merimal Flinitt, and Brandee Dudeck.
- We're planning to use the money to expand our team, particularly in engineering.
- Year to date, GutenberGo's user base has grown 14x, and it's only picking up steam.

*Post (we did a quick call before I wrote it, so there's a little additional information included):*

**PC Load Letter?! GutenberGo Raises $1.4 Million So You Never Have To Touch A Printer Again**

GutenberGo, a San Francisco-based startup founded last year, has raised a $1.4 million funding round led by Platitude VC and several angels.

Founder Joanna Sagan says that GutenberGo will use the money to expand its engineering team (the team of two will grow to half a dozen, and will be working on apps "for all the major smartphone platforms.")...

### Growth and other numbers

Congratulations, you've passed an arbitrarily large milestone! Generations of startups have abused this story-type so reporters are on the lookout for 'vanity metrics' (you'll be hearing more about these). But concrete stats can still make for a great story, and plenty of vanity gets through.

*Pitch:*

### GutenberGo prints 10,000 pages in eight weeks (want the exclusive?)

Hey Jason,

Wanted to give you a heads up on some of the engagement numbers we've been seeing lately that we're really excited about. We'd love to give the story to you exclusively — please let me know if you're interested! Here's the gist:

- After launching GutenberGo eight weeks ago, we've just printed our 10,000th page for our customers.
- Growth has been incredible. We're seeing 150% weekly growth in new signups, and existing customers are using the service *more* over time (it's easy to get used to not having to deal with printers!)
- We just updated our iPhone app with real-time courier tracking, and our Android app will be coming out very shortly (I'll give you a heads up when it is).

Let me know if you have any questions, and I'm free any time to hop on a call.

Joanna

*Post:*

### GutenberGo's Printers Press Their 10,000th Page, Growing 150% A Week

Hot on the heels of raising $1.4 million from Platitude VC and others, 'Uber for printing' startup Gutenber-Go has just printed its 10,000th page. The company has also seen strong growth, with triple-digit percentage increases each week…

Founder Joanna Sagan declined to specify exactly how many *documents* those 10,000 pages represent (which is perhaps a more telling indicator of how it's being used), but says that "the vast majority of documents are ten pages or less"*. She adds that once a user has signed up for the service, after an initial experimentation stage they often begin using it on a weekly basis…

### Partnerships and integrations

This category gives birth to heaps of bad pitches. PR people like to trumpet about how their company has just launched an integration with so-and-so other company, neither of which are popular.

When you do a partnership or new API integration or whatever, don't spit out a press announcement because that's what you think you're supposed to do. You've been mainlining your own Kool-Aid — just because you're excited doesn't mean anyone

---

* See what she did there? It's possible that one guy keeps printing 100+ page documents, which would make this 10,000 stat totally misleading. But hey, if I don't write this story, someone else will…

else will be, and you don't want to over-pitch. Consider posting it on your blog, or tuck it behind year ear til we get a few paragraphs further in this chapter.

*Borderline pitch:*

> Hey Jason,
>
> Things are going great at GutenberGo — wanted to let you know that we've just integrated support for Quadrangle's location API. This will let us better handle orders from public locations, like concert venues, making it that much easier to get your last-minute tickets delivered to you while you're at the show…

**Post?:**

This pitch is iffy. If it were a slow news day then maybe I'd bite (I mean, it *does* sound handy), but Joanna would be wise to flesh it out with a lot of supplementary context — say, some stats demonstrating how often GutenberGo users are ordering tickets from concerts.

### New features and other improvements

Post-launch you'll implement a bunch of stuff you would've loved to ship with but had to push to version 1.1 — plus you'll see how people are actually using the thing. Several weeks of all-nighters later, your app will look like post-spinach Popeye.

*Streamlined signup flows. Additional configuration options. A classy, nighttime color palette called* Dusk.

But reporters won't care, not unless you sell it the right way. Because *everyone* is always adding new, cool stuff. It's one of the virtues of this industry, but it means that incremental updates are hard to get excited about. And it isn't just a matter of meeting my nebulous newsiness quotient. Sometimes there just isn't much to say.

Imagine you're writing an obligatory *'Thank You'* note for a gift you didn't want, from a person you hardly know — "Dear Cynthia, thank you so much for the Google toaster oven..." — and you get one paragraph in before realizing your mental inkwell just hit empty. The card's still uncomfortably bare, so now you get to spend a relative eternity agonizing through two more paragraphs of fluff ("it says my name aloud when my toast is done!").

Now imagine doing that for an audience of thousands and tell me *you* wouldn't pass on writing a post about PixelYak's upgraded storage limit.

This brings us to the concept of bundling, which will play a key part of your post-launch press strategy.

Say GutenberGo has been working on several new features and is figuring out how to announce them. They've:

- Added support for the latest-and-greatest digital currency.
- Cut the number of taps required to order a print job by 30%.
- Added a new synthetic fiber, eco-conscious 'faper' option.

If Joanna were to pitch me on any of those bullet points on their own, I'd probably pass. But there's strength in numbers. If we tie these bullets together in one pitch — bundling, as it were — we start to approach enough content for a post. To flesh things out we can pull in additional points from the other categories, like milestones and integrations.

Some pitches just toss a few of these bullets in the same email and call it a day — if you ever see a story that reads, "So-and-so does this, that, and the other thing", that's what happened. But why not call on our old friend *story* to make this a little more cohesive?

*Pitch:*

### (exclusive) GutenberGo amps up War on Printers by declaring War on Paper

Hey Jason, thanks again for covering our funding news last month. Wanted to follow up with some of the improvements we've been making, and how we're expanding our battle with printers to take on paper, too. We'd love to give you the exclusive on this — let me know if you're interested!

- The average American has three reams of paper in their house and another five in the office. This represents half a tree *per person.*
- With GutenberGo, all these people can stop stocking up on paper (and printers), using only what they need.
- We're introducing a new synthetic "faper" option, so eco-conscious customers can avoid killing trees entirely.
- We just added Bitcoin integration, and for the next 6 months are donating 3% of payments made via Bitcoin to Greenpeace.
- We've made many other improvements to the app, including reducing the number of taps needed to place an order by 30%.

Joanna

*Post:*

**GutenberGo Saves The Environment and Your Sanity at the Same Time**

GutenberGo, the 'Uber for Printers' app that launched early this year, has just unveiled a new initiative it's calling *The War on Paper*. It's bizarre, sure, but listen to founder Joanna Sagan describe it and it sounds like the quirky startup might be onto something.

"We're already well on the way to mitigating the soul-crushing experiences that are modern printers, and we realized we could take out two birds with one stone," Sagan says. Next on her list: Global Warming.

# Angles

Breaking: Google is launching a landmark advance in energy efficiency that will reduce the carbon footprint of its servers by 40%. A dozen intrepid reporters are dismayed to see they've been scooped by one of their rivals. They still need to write something — it's big — but what good is parroting the same thing again? Surely they can do better, something that adds to the conversation (and justifies their job). Better think of an angle.

And so they run off and write something hyperbolic ( *"Google's Gambit to Save the Planet"*), something contrarian ( *"Google's Green Smokescreen"*), or something out of left field ( *"Is This an Allusion to Nuclear Fusion?"*). Do justice, and let the page views soar.

Angles are story lines for news, where the plot diverges from the standard "who what where when". The reporter isn't just writing how Google is launching this environmental thing, they're weaving a story that paints how it could be wonderful or terrible or whatever. Herein lies an opportunity.

Historically reporters have been taught not to express their opinions in stories because they introduce bias. So, when they want to go beyond 'just the facts', they turn to sources and industry experts, gathering quotes and data they can weave together to craft their story. Of course, the reporter still chooses *whom* they talk to (experts are often predictable) and which quotes to include (two sentences, snipped from a twenty-minute conversation), but at least it looks the part*.

Here's your chance: you, too, can be part of a story that you're not really part of, by helping these reporters find an angle.

To do this, you need to contribute to a story that's bigger than you, without making it look like your pitch is *about* you (but, really, it is). Good angles often involve data you uniquely have access to, which could include previously-unreleased figures observed in your app around user engagement and habits. Say: Instagram launches animated stickers, you're an animated sticker app, you have numbers on how sticker-use has been on the rise over the last six months, particularly for Android users. (This is also your chance to paint it as an "all-ships-rising" story, as opposed to the more obvious "Instagram-is-about-to-crush-you").

Many angle pitches revolve around broad trends, so you can take your time. But with the example above — Google's break-

---

* The rise of blogs has driven more reporters to openly express their dirty, honest opinions, but they still like to buttress their thoughts with outside citations.

ing green-thing — we're dealing with a fast-paced version of the angle I'll call the Hop-on.

**Hop-ons**

In the wake of big news, PR people spit out countless angle spermatoza in the hopes of breaking into and becoming part of the main story. They know that reporters are trying to find a unique take — and they have just the data-point/expert/rebuttal to make that happen, right now. You can do this too.

Timing is key. Pitch too soon and you risk getting lost in the deluge of similar attempts; too long and you're hawking stale news. I'd wait a day, maybe two, while the news is recent but has enough of a gap that your pitch won't be overwhelmed by the original story you're trying to piggy-back on.

Let's consider how GutenberGo might approach this Google news.

Joanna realizes that on-demand printing has roughly zilch to do with green servers — so sharing proprietary data won't do much good here — but spots a glimmer of opportunity stemming from their newly-launched *War on Paper*. She quickly contracts a designer to make an infographic contrasting Google's launch with green initiatives led by other companies (*"Who Saves More Trees than the Big G?"*).

GutenberGo's own efforts are nascent and don't belong in the infographic itself, but it'll include a small sidebar inviting readers to use GutenberGo if they, too, would like to help the environment. Take note: if whatever you're touting is obnoxiously self-promotional then reporters will likely pass, but you can get away with a little bit, and they'll probably link to you in their post.

Joanna pitches the infographic to reporters and mentions she's available for interviews on the topic (the press doesn't have a high bar for what constitutes an 'expert'; the fact that GutenberGo is waging a *War on Paper*, regardless of Joanna's prior eco-credentials, may be enough to convince a reporter that she's worth quoting). With any luck someone will bite, they'll do an interview, and forty minutes later we'll be reading about GutenberGo and Google in the same breath.

For more ideas, look for news stories where a small company's data and anecdotes are discussed in relation to a big story with an only-tangental connection. Incidentally I think — *hope* — that once-ubiquitous infographics are losing steam, but it will only be so long before they resurface as digital experiences with touch-based interactivity and dynamic soundtracks, god help us.

Finally: it looks desperate when you keep trying to tie yourself to big stories to which you are completely irrelevant, and I will curse you forever if the subject line for your mediocre infographic pitch reads, *"Breaking Google News!"*.

# Politics

We've established that exclusives are a good way to build relationships with reporters. Joanna uses this to great effect throughout this book and builds a strong rapport with yours truly, handing me exclusive after exclusive.

This is a bad idea.

I've written it this way because it's an issue most startups aren't lucky enough to be concerned with; many founders would kill to have reporters get jealous over them.

Indeed, in the early days, having a single reporter that believes in you — and will cover you regularly — can be the difference between obscurity and serving as the subject of resentful conversations across San Francisco.

But if your company has luck and legs, this kind of favoritism is a good way to get other reporters gunning for your kneecaps. You're going to have to learn how to make everyone happy.

Let's revisit the relevant issues:

- Reporters don't like sharing (via embargoes).
- Reporters loathe when their competitors are granted exclusive access to a hot new product. Sometimes they even get jealous over exclusives they would have ignored had they been offered to them in the first place.

These two points are obviously at odds; you're always going to be pissing someone off. It's a matter of degree. If I'm covering a cool new app and two other reporters are doing the same under embargo, I'll just shrug and write the post, the way I've done a thousand times before. But when a hot new app or a company I've previously covered doesn't give me so much as a teaser? Fumes.

The resulting solution is a thankless dance of ego massage. Fortunately it's one the big tech companies have honed into an art, and since many of the steps are readily apparent in the daily news cycle, we can follow their lead.

*Step one:* Save embargoes for your big stories. The bigger a story is, the less reporters care about sharing (I'd rather you just give it to me, but better to be in on an embargo than left out in the cold).

*Step two:* Contact your favorite reporters at regular intervals with scoops they can have exclusively, which boost the odds they'll generally like you (and be willing to positively cover those big embargoed stories). These often take the form of controlled leaks, where your involvement isn't necessarily apparent to other reporters (though they may have their suspicions). Other common favor-enhancers include:

- 'Sneak peeks' at upcoming products (while reserving a full review for an embargoed launch).
- Offers of interviews with key personnel.
- A behind-the-scenes look at anything from your convention-defying development process to how your shrubbery has been hand-crafted to promote Feng Shui. Even better, these often read like puff pieces (but you'll need something nifty to show off).

The trick is to ensure these scoops are newsy enough for the reporter to feel good about writing them, but aren't big enough to inspire too much frustration in the reporters you didn't give the story to (and: you need to make sure they get theirs soon, too).

Another key tactic is to make embargoed news *feel* special to each reporter, even if only superficially. You can do this by reserving an exclusive tidbit for each outlet: "We're only sharing this number with you..." or "We'd love to set up an exclusive interview with the app's product manager..." should do the trick.

You're still going to have to deal with annoyed reporters (even the best PR people have this issue; finessing relationships is one of their most important skills). But remember: it's a good problem to have.

# Competitor Writeup

Your phone vibrates. It's dark. It's been, what — two hours — since you let yourself close your eyes, but there's no time to feel miserable because your heart is pounding as you parse the text message from your cofounder.

"fuck." and a blue link.

Your competitor is at the top of *TechCrunch*. Rave review. You've been trying for *months* to get TC to respond to a single email, and now this lame startup whose product could barely be considered pre-alpha convinced this writer that they're the second coming. The fury rises.

It happens to someone every day. You'll need to fight your first instinct, and probably your second. Swear loudly, complain to your cofounders and investors, but do not under any circumstances send an email to the reporter whining about the injustice of it all, or make a desperate plea to get them write about you. This is not the time for a hop-on.

We're going to approach this from a couple of scenarios.

The first is the default: you haven't received press coverage before, and now this competitor of yours has managed to make themselves look like they have a first-mover advantage, whether they've been around longer than you or not.

The amateur move is to send an email to the reporter immediately after the story goes live that obliquely questions their integrity ("I've been pitching this exact thing to you for months, guess I should have done Y Combinator..." is not going to score you any brownie points). People do this. Dumb.

Another common tactic is to wait a few days and attempt to pitch what is effectively the same story. Even if you do a bang-up job writing that pitch, in the reporter's eyes it feels like you're just saying, "Hey, we do this too!" — and it's no fun to write the same thing twice.

A better strategy: wait a couple of weeks so the slog of writing your competitor's story isn't so fresh in the reporter's mind, then hone in on that magic phrase: *how are you better or different?* That's your pitch — why your mission is focused in this special way, and how you've nailed this feature because of it.

There are two reasons why this is a glass-half-full situation. The first is that the reporter already has a sense of context — they have a basic understanding of the problem you're solving, even if they learned about it from a competitor. And second: there's an inherent source of conflict, which is always fun to write about.

You don't want to bash the startup that got written up — it's easy to come off as a bully. Instead, try a quick mention of them early in the pitch to establish context before moving on to phrases like "No one else does this." and "We're uniquely targeting X, our competitors are all doing Y". The conflict is not you versus anyone in particular, but rather you versus everyone. The reporter may still pit you two head-to-head in their post (maybe that's what you want), but better for it to be in their words, not yours.

Onto scenario two, which differs from the above by a single email.

If you *have* gotten significant press coverage before, particularly from the same outlet that wrote about your competitor, and the story includes no reference to you, then you can consider emailing the reporter shortly after their post goes live and asking for "a mention".

To state the mind-numbingly obvious: a comprehensive post about a product should include a discussion of its competitors. Strange, then, that, reporters will often either omit this discussion entirely or present only a subset of the relevant competition. Sometimes it seems they are totally oblivious to the fact that they exist at all.

Which they are. Not always, of course, but more often than you'd think.

Because when your beat is something so broad as 'startups' or 'tech', you wind up writing about dozens of industries, and it's impossible to keep tabs on the state of everything. Even the stuff you do know blurs together (no thanks to over-use of the 'pivot' strategy). Reporters aren't *trying* to forget you. They just do.

Which is why it's acceptable in some cases to give them a gentle nudge and say, "Hey, remember us?"

Like this:

"Hey Jason, saw the article on PrintDriverz this

morning. They're doing some cool stuff — I wanted to ping you because we've actually been doing a lot of the same things for a while. We'd really appreciate if you could mention us in this context (your colleague Mark Hendrickson wrote about us here), and either way I'll shoot you a note when we have something cool to announce."

For me, the reporter who wrote about PrintDriverz, this is an easy, "sure". It'll take me 30 seconds to add a link to your start-up, and the fact that you've already received coverage gives you credibility.

The trouble with these requests — and the reason I prefaced this scenario by saying you've gotten press coverage previous-ly — is that reporters are often inundated by similar asks from companies they've never heard of. Which means they need to research whether you're a viable competitor or just a wanna-be, which will take time, which doesn't bode well for you because there are roughly a million things that are higher-priority. Proba-bly better for you to wait a few weeks, and GOTO scenario one.

# Ye Olde Secret PR Handbooke

There exists a secret guide passed between generations of PR people that details their secrets to success. I have not not seen said guide — it is doubtless kept under lock and key, where new inductees are told to commit it to memory — but nothing else could explain the consistency with which they work.

Herein follows the advice I have been able to derive based on extensive research, along with my own commentary.

∾

### Thou shalt compliment thy reporter's work

*Reporters really like it when you say nice things about their writing. After a long slog through mean comments and impersonal pitches, it's a breath of fresh air when they see that someone appreciated their latest thought-piece or major news-break.*

*So do your homework. Read their recent articles, and compliment them on the pieces you liked best. This is a good way to lead into your pitch, as it shows the reporter that they are not simply another recipient of a form-filled email.*

This piece of advice seems sensible enough — and, indeed, reporters see its fruits every day — but it makes the mistake of assuming that reporters will perceive these compliments as if they are genuine. They do not. Reporters have been brown-nosed so many times, by so many people, that any attempt at flattery, no matter how sincere, likely means they'll be rolling their eyes as they skim the remainder of the pitch (eye-rolling has a decidedly negative impact on reading comprehension)

I'll hedge that a bit. If the reporter recently wrote something that is directly related to whatever you're pitching, or you have extensive experience in the industry they wrote about, or you know them personally, then you can mention it. Just don't give a compliment because it's the first thing on your checklist.

On a related note, skip the "I know you're really busy but I think you'll like this…" prelude. I am busy, so let's get to the pitch, shall we?

### Thou shalt use prettie and unique fonts

*Every email message tends to look alike — the same font, the same colour, even the same spacing. It is our job to make our clients stand out from the rest of the pack, and making your email less visually generic goes a long way.*

*So experiment! Green, blue, the sky is the limit (be careful with sky-blue, though, as it can be difficult to read). Dabble with smaller fonts, try out something bigger and a little cursivey — and don't be afraid to mix and match. Don't overdo it (you don't want your pitch to look like a Hallmark card), but remember: drab is bad.*

While it is true that most emails look alike, this isn't a bad thing — reporters get really good at finding the important bits. Sending a message laden with visual flair (even if it's just an overuse of bold and italics) can be distracting in a way that hurts the pitch.

To be fair, there are some PR people whose emails I've recognized based on minor deviances from the norm. Some of them make their fonts a little smaller, some go with dark blue, some use Garamond instead of Arial — but these people are good at their jobs. They're not good *because* of their font choices, but rather because they send good pitches, don't have dumb responses to questions, and generally treat me like a human being. Their fonts help me recognize them, but they don't make me like them any more.

Don't break this rule until you're good enough to break it.

### Thou shalt tell fibs about broken email and phone lines

*No reporter likes to be hounded about responding to a pitch — but you still have a job to do. If they don't respond to your first email, try again. If they still haven't responded, it's time to pick up the phone.*

*But you musn't ask them directly whether they are interested in the pitch, for their answer will almost certainly be in the negative. Instead, pretend as if you've been having email troubles (what with technology being what it is these days), and just want to make sure your messages have reached their destination.*

Either the entire PR industry has fallen prey to unusually terrible email software, or they've all bizarrely decided that this excuse — roughly as believable as the canine homework-muncher — passes muster. If you're calling to pester a reporter about a pitch, you may as well be honest about it (to be clear, cold-calling in the first place is an aggressive tactic that's liable to piss off whoever is on the other end of the line).

### *Thou shalt engage in social media minutiae, then immediately send thine pitch*

*Grabbing the reporter's attention is the name of the game, so use all the tools you have at your disposal. Get social: Tweet, Facebook Message, Poke — anything that makes thine name top-of-mind, then send the goodes!*

Sometimes I'll respond to someone on Twitter without realizing they are in PR or working on a startup and immediately get a pitch from them. Ugh. It's not the end of the world, and yeah, I guess I'm more likely to read it, but I'm tasting bile as I do.

∾

# Sonorus

This isn't a book on managing your social media strategy, but there are many others that will tell you how to tweet, blog, and inspire moments of heartfelt connection with the good people our industry affectionately refers to as "users".

I will, however, take a moment to address some issues relevant to your dealings with the press.

First: do not break your own news before a reporter does.

Companies do this. They'll brief a reporter on their launch or some new feature, and then they'll tweet about it or publish their announcement to their own blog before it hits the presses. I think this stems from the fact that most people understandably don't care "who's first" to publishing a story, but reporters definitely do care — it's *everything* — so don't muck things up.

Before you get too smug ("I'd never do *that*"), you should know there are myriad ways someone on your team might break your news for you. Like the employee who posts a teasey, "Wait til you see what we're doing tomorrow...", which isn't so terrible, but then a reporter sees it and does some digging and their post — which is 90% wrong — blows up the embargo you've been working so hard on.

You can't prevent every bullet, but make sure your guy isn't the one pulling the trigger.

Next: figure out a blogging strategy. Blogs give you a way to get your message out without having to go through the press. This is handy when you need to release a timely statement (say, in a crisis situation), and it's also good for maintaining your drumbeat.

As I've said elsewhere, one of best ways to damage your relationship with a reporter is by inundating them with pitches for stuff you're doing that isn't newsworthy. But that doesn't mean it isn't worth sharing somewhere. Your blog is a great place to share anecdotes about how people are using your product, incremental updates that'll make existing users happy, and other 'small' stuff that still seems pretty cool to you — especially if it serves to support your story.

And if you pitch reporters on something and no one bites, write up what your ideal post would have looked like and stick it on your blog, then post it to Hacker News and wherever else makes sense. Maybe it'll get people talking anyway.

All of this comes with a caveat: anything you've already published — on Twitter, on your blog, whatever — is unlikely to add to the 'newsiness' of a pitch to reporters down the line. The reporter may be happy to include this already-announced information to flesh out their piece, but they won't be writing that article *because* of it*. To that end, make sure your pitches are unambiguous as to what has been previously announced. Few things inspire reporters' bloodlust like getting hoodwinked into covering old news.

Finally, if you get press coverage, you can tout it on your own blog and include additional details the article(s) may have glossed over. Be sure to prominently feature a link to their piece, and give it some lead time — you're supporting them, not competing with them.

The options for publishing are evolving. For years startups have set up their own WordPress or Tumblr blogs and posted to them occasionally, but these entail annoying overhead; let yours go stagnant and it reflects badly on the company. As I write this more startups are turning to Medium, which lets you publish posts in a one-off fashion so frequency isn't a concern (but: you lose control over the experience). The jury's still out. Look at what your neighbors are doing.

---

* If your pitch revolves around a broader story you've documented extensively on your blog (say, GutenberGo does monthly blog posts about the 'War on Paper') then these posts can serve as supporting evidence.

# Preempting the Lightbulb

All startups have a weakness, some threat that spells out their likely cause of death long before they even launch. In many cases it's obvious: the chicken-and-egg problem that renders the social app a ghost town; the license fees that bring the music app to its knees. For all their diversity, startup death spirals follow predictable patterns, and most of them fall, eventually.

Which is why it's a bad idea for you to pretend that you have nothing to worry about.

Founders do it all the time. They'll present their pitch, glossy and carefree, without giving the faintest indication that anything they've discussed will be difficult in the slightest. They've doubtless obsessed about their pitfalls to no end and don't want to reveal how worried they are, but they make the facade *too* perfect.

Sometimes it works — lazy writers won't notice. But a good reporter will see through this and ask the tough questions, no matter how convincing your grin is.

Which is fine. Reporters know everyone has problems to deal with. They'll ask, and you'll give your sanguine line about getting distribution via inherent network effects, and they'll include some skepticism in their article. But what else can you do?

Here's a trick that makes the unavoidable a little less bad: tell the reporter about your issues before they ask.

Bear with me. For reasons related to *All the President's Men*, there is a tiny lightbulb that goes off inside the reporter's mind whenever they stumble across something they think you don't want them to ask about. It's what *real* reporting feels like, and *real* reporting inevitably leads to wiggling eyebrows and skepticism. So beat them to the punch.

As you talk about your product, point out the obvious obstacles you'll need to overcome and how you're going to do it. I'm not saying to ring your death knell — keep it vague and upbeat. You'll be conveying the same overly-optimistic information you would be if the reporter had asked about it, but this way you're framing the issues on your terms, making it sound like you know what you're doing (which includes knowing where your challenges lie).

It won't solve your chicken-and-egg problem. But anything to keep that lightbulb from going off.

# Press Releases

Press releases are relics from an era when companies announced their news via wire services. They still do, actually, but I've never heard a reporter refer to them with anything more pleasant than bemused contempt.

Your standard press release reads like a coked-up narcissist's dating profile. It boasts exuberantly about everything, executives trumpeting their superlative-laden excitement from the hilltops (meanwhile, everyone knows the words never left their mouths — PR people write press release quotes*).

What's worse: press releases obfuscate the actual news, burying it in a pile of noise. Reporters can sift through them quickly

---

* If you see one of these quotes in a news article, it often means the writer was lazy and looking to boost their word count.

— see enough and you develop a sort of X-ray vision for picking out the nuggets of substance — but why not save us the trouble?

Most reporters much prefer email, which allows for directly targeted and concise pitches. That said, sometimes PR people will include a press release as an attachment to their email pitch, alongside screenshots and other supplementary material. I guess the logic is that some reporters cut their teeth on them, so they like the familiarity, or maybe there's a 'long tail' of bloggers who subscribe to the news wires and crank out posts that get a couple hundred hits. I don't know.

If you're managing your own PR, don't bother with this — you have enough to worry about[†].

---

[†] Okay, fine. PR people I've spoken to say press releases can have *some* benefit in that they can do well in search results (better to have Google link to a puff-piece you wrote yourself than to something negative or unrelated). It's still the last thing I'd be worried about.

# Fun With Numbers

Numbers are reporter catnip. They make us a little stupid.

From app downloads to funding rounds, percentage growth to acquisitions — the industry is buoyed by a raft of zeroes that confers a not-so-rational exuberance to the whole thing. A million downloads?! *Surely this calls for cake!*

You can use this to your advantage, because god knows your competitors will try to.

Despite the press's directive to dig up things they aren't supposed to know, the majority of numbers in the news come straight from company mouthpieces. Reporters have roundabout ways to discern things like traffic, but these rely on third-party data sources whose methods are suspect and can easily be shrugged off. In other words, tech companies get to choose which numbers they want to tout and when — which means you can cherry-pick the ones that make you sound good.

Let's start with the number you probably won't give away.

*Active Users* is a simple measure of how many people are using your product within a specific timeframe, say, "ten thousand users in the last month". Given the myriad numbers flying around it seems a reasonable one to inquire about — what good are the others if you don't know how many people are actually *using* the thing? — but almost no one shares it, because it is the truth.

Say you're a startup and things are going pretty well. In the last month you've had 52,300 people log into the app, and the growth chart is pointing up and to the right. A reporter comes knocking and wants to get a sense for just how hot you are. "Real hot", you'll say, and you'll reel off some impressive figures — none of which will be 52,300.

For starters, you're nervous about the competition: you don't want them to get a sense for just how well things are going (it gives them a chance to definitively show that they're performing better than you). And — wisely — you don't want to set a precedent. Things might be going great now, but what happens in three months when you're launching a revamped iPhone app, and the count has dipped to 40,500 monthly active users? You could say you've decided to stop sharing that figure, but that's going to raise eyebrows — it'd be much easier if you'd kept your figures vague from the get-go.

Obviously there's a trade-off here, because firmer data instills credibility, and reporters are jaded by the constant barrage of deceptive stats they face. But: big numbers make for enticing headlines (it's that booze kicking in again), so the bar isn't *that* high.

Here are a few ideas for 'newsy' numbers that only reveal what you want them to.

### Total users/downloads

The cumulative number of people who have ever signed up for your service. This is a handy number because it never goes down; so long as your app has done reasonably well at some point in its history, you have something to boast about. This is the most-commonly abused of vanity metrics, and while some reporters have taken to quashing it, it still pops up regularly.

### Number of X sent (per Y) (by people who do Z)

Common variations on this include "total photos uploaded per user", "total messages sent (and/or received) by users with more than 3 friends" and so on. Depending on the app it's possible to massage these so they imply much more activity than reality would suggest, particularly when a small subset of die-hard users is skewing the average.

### Percentage growth

When you're getting started and don't have many users, it's not uncommon to see user and engagement growth in the double or triple digit percentages. Any 8th-grade algebra student could tell you that these percentages are meaningless without a sense of your start point — 80% more than *what?* — but reporters aren't so concerned about this.

*Pro-tip:* When you choose how to bound this stat (in the last month, in the last three months, etc.), keep an eye out for an unusually slow period to use as your 'before' point — your subsequent vitality will seem that much more impressive.

### Restricting metrics by region or demographic

So your app isn't doing so great in most of the world, but for some reason it's really taking off in Mexico. By all means, do some analysis that scopes the data to this hot pocket. Have a couple of anecdotes or trends ready to explain the increased activity there, even if the causation isn't exactly rock-solid (e.g. "It's really taking off with high schoolers via word of mouth").

### Creative definitions of words like 'active' and 'user'

Some companies will define a user as anyone who stumbled across their website, or saw an embed, whether they were logged in (or even had an account) or not. This can lead to staggeringly large — and completely misleading — user figures. Be warned: if a reporter catches you doing this, they will not respond kindly.

### Studies and infographics

Some companies get really good at compiling "Reports" describing one trend or another. A lot of these stem from unconvincing datasets or have clear bias issues, but so long as your conclusions sound newsy and you assert them with confidence, plenty of reporters will be happy to print them anyway (they might discuss the caveats, but not til the seventh paragraph).

There are a million ways to slice your numbers, but they do you no good if the reporter won't print them. There's no easy rule here — keep an eye on what they've gone for in the past — but generally speaking the more central the number is to your story, the more concrete it needs to be. If you're trying to pitch how much you've grown since launch, you'll need to cough something up that makes it clear that people are actually using the product. Maybe that isn't active users — GutenberGo might get away with a pitch on printing their 25,000th document — but it can't be pure fluff.

If the pitch revolves around something *other* than stats — say, you've landed a new funding round or are launching some com-pelling new features — that's the time to get more creative. The reporter may ask for more, but they'll probably still print what-ever you give them.

*Cake!*

# Crisis

Let's hope it doesn't come to this, but eventually you may have what PR people would call a Crisis Situation — shit's hitting the fan.

If there's ever a time when having a professional on your side is worth it, this is it. No guide is going to help you when a hacker uploads a dump of user data to the web and a dozen reporters come pounding on your door. On the bright side, if people care enough for you to have a crisis, there's a decent chance you're big enough to be paying someone to handle this sort of thing.

But you're here, so I'll take a stab. I've never been on the PR side of a crisis, but have written about plenty. Here's what I've learned:

**Be prepared.**

Most crisis stories revolve around familiar themes, so you can do work ahead of time to keep yourself from getting caught on your heels.

- Security breaches are common enough that you may as well have an outline prepared. If you haven't found any evidence that sensitive data, like user payment information, was accessed, be sure to say so*.

- If you're doing something shady, don't be surprised when you get caught (I hope you get caught). As you consider potential sources of shadiness, remember that as a founder you're exceptionally good at rationalizing things that keep you afloat. Also: the fact that you're aware of other companies (even big ones) that are doing the same shady stuff does not indemnify you. Hey, maybe you could send a reporter a tip about them…

- If you're even discussing layoffs, know that they may leak early. Be sure you're clear on what you're allowed to say regarding HR issues.

- Have you or any other executives ever said something stupid, particularly in an environment where it may have been recorded? Don't assume it won't bubble up later. Scandal has a long shelf-life.

---

* Companies will often proclaim that there is "no evidence" that precious user data was accessed in a security breach. In turn, the press nods its collective head — "you heard the guy, no evidence!" If there were anyone ill-suited to detect this evidence, you'd think it'd be the folks with the swiss-cheese security, but what do I know?

### Get control of the story.

Eventually you'll want to be as direct and candid as possible — but it may take you a while before you can write a blog post or prepare a statement that lives up to that goal. The internet isn't going to wait, so in the mean time, get in touch with a reporter you trust and ask to speak to them *On Background* (or *Off the Record*) so you can fill them in on the current status; they can work this information into their story without presenting it as your statement. If you're still looking into the issue, tell them so. If you're not quite done with the statement, they'll probably understand. Unless, of course, you were doing something shady.

### Don't try to weasel your way out.

Own up to what you've done wrong, explain why it happened and how you're going to make things right (or at least, ensure it doesn't happen again) — and don't use a technicality to explain why you're not *really* at fault. If news stories have mischaracterized the nature of the story to the point that it could damage you even worse then the truth (e.g. saying a breach was worse than it was) then you should provide the facts to clear the air — but don't go on a rant about media sensationalism. Even an apt critique will seem overly defensive (this is a common impulse that never ends well).

### "A small number of users was affected."

Take the number of people affected by your security breach (or whatever it is) and divide it by how many people use the app (or if you want to be really deceptive, have *ever* used the app). Is it a smallish-sounding percentage? Great! Include that in your statement (or just use the quote above). Never mind that this could represent a lot of people — companies like Google and Facebook do it, and their "small number" could be a hundred thousand, so you're practically a saint[†].

**Sometimes words aren't enough.**

Heads roll nicely. Probably best to wait awhile, though; this is more penance than short-term damage control.

**Get a second opinion.**

If you don't have a PR person, at least talk to someone with good judgement who can give you objective input on what's happening, and how your work-in-progress response sounds. Don't trust yourself when you're in panic mode.

**Really, you should probably hire someone.**

---

[†] To be clear, you should never give the impression that it's acceptable to have *anyone* affected. But companies will often present the softest-sounding version of the truth.

# It's Technical

It's counterintuitive, but many 'tech' companies have very little to do with advances in technology; often they're just twisting existing tools in different ways. This isn't a bad thing — it dramatically lowers how much money and time are needed to build something — but it enables a problem: many tech reporters aren't exactly experts in... tech.

Usually, they don't need to be. But every so often they'll write a story that betrays them: a sensationalized post about hacking, a fundamental misunderstanding of how APIs work; most programmers don't have to think hard to remember a 'tech' story that made no sense. Which of course presents a problem if the company you're building *does* involve genuine technical innovation, or is catering to an audience of programmers and engineers.

Let's start with the easy issues. Say you're building a social app that uses complex algorithms to suggest new friends to people. Whatever you do, don't explain the math involved, no matter how elegant it is. Instead, focus on describing the result of these algorithms, and why you're credible enough for the reporter to believe (or at least print) your claims. Say your pitch includes a couple of bullets like this:

- We use novel genetic algorithms to pair friends. Early beta tests show that these are 35% more accurate than traditional models.
- Our algorithms were developed by Stanford engineering Ph.D. Vint Bool, who has been conducting research in the field for six years.

And — *voila!* — here's what goes in the post:

The app uses genetic algorithms developed by Stanford Ph.D. Vint Bool, who has been researching the field for six years, to match friends with each other. Bool says that early tests show they're 35% better than traditional models.

Maybe the writer reworks it to be less verbatim, but the point is this: when you're dealing with stuff that reporters don't understand (or have any way to verify) there's a good chance they'll print whatever you tell them, provided you have the credibility to back it up.

A much harder challenge is getting a reporter to care about a product that is inherently technical, with benefits that only make sense if they understand the tech in the first place. As an example, let's consider a company that's building a new kind of database.

Their pitch kicks off:

> Hey Jason, wanted to reach out to you about a new
> kind of database we're building called TinyChair. It
> allows for seven values to be stored at any given time.
> We're planning to release it next week with support
> for NoQuery...

— woo boy, you've lost me.

The problem is that I have absolutely no idea what you're talking about. I mean, I sort of do, to the extent that I can define some of the words in the pitch, but I have zero sense as to what the context is, which means I have no way to figure out whether it's a big deal. I'm intimidated so I move on to the next email with a vague intention to circle back to this one later, which means it never gets seen again. This is suboptimal.

You have a couple of options here. First, you could stick with reporters who actually understand databases, programming languages, and other engineery things. There are some, but they're in the minority, mostly because they can make a lot more money as engineers.

Your second option: rework your pitch so that it revolves around layman-understandable points, with a few bullets that get into the nitty-gritty tech stuff that the reporter can just quote verbatim. Which means we need to come up with some easily-understood use-cases. You don't need many — just an anecdote or two that anyone could understand. In other words, it's story time.

> Hey Jason, wanted to reach out to you about a new
> kind of database we're launching next week called
> TinyChair. It has tons of potential uses, but so far
> the coolest one involves artificial intelligence — AI!

Here's the gist, please let me know if you're interested and I'll fill you in on all the details:

- People get creeped out by AI because it remembers *everything*.

- TinyChair is super-fast but can only remember seven things at a time, which is roughly analogous to human working memory.

- Early results with robots equipped with TinyChair have been very positive — Celiceo Hammerling, CEO of BigBots, recently told us, "Nobody likes to feel dumb, especially when they're talking to a machine. TinyChair helps us strike the perfect balance."

- A dozen other big-name companies have shown interest in implementing TinyChair.

- We use *(this is where you put all the tech jargon I'll just copy and paste because I don't know what it means)* a schemaless setup with JSON for its DBMS, resulting in query times that approach $3.0 \times 10^8$ pulls per second.

# Don Draper

If I'm a PR person, one of my knocks against this guide is that *real* PR can't be captured in a book because it stems from opportunistic creativity. And it's true: many of the best campaigns involve ingenuity — not to mention a considerable amount of risk-taking. They're stunts.

Reporters love these things despite themselves. A successful PR stunt won't just get coverage on the blogs, it'll go wild on social media, which means more eyeballs pointed in your direction (and page views for whoever writes about it). Some stunts, like Airbnb's cereal boxes*, are the stuff of startup legend. Most are quickly forgotten, but you don't need to make the history books for a stunt to pay dividends.

I can't tell you how to have a creative idea. But I can tell you to be careful. The biggest PR stunts are the ones that go wrong.

Most fizzle; no harm, no foul. But in order for a stunt to reel in the bloggers and social-sharing masses, it has to be extraordinary. When there's a different startup pulling shenanigans every week, that becomes hard, leading to a sort of one-upmanship that walks a fine line between cheeky and people waving pitchforks. The snark-makers of the world know that tone-deaf marketing stunts are traffic goldmines and will not hesitate to find ignorance or bigotry in your intent (sometimes, they're right).

The good news is that you don't need to bother with stunts if they aren't your thing, particularly early on. They're the exception to the rule, which is why they get noticed at all, and a lot of stunts are from companies with the money and staff and brands to make their bold ideas significantly less risky.

My sense is this: if you've got an idea you can't kick, some perfect ruse that'll make people laugh and tell their friends about you, then run it by a few trusted people for a sanity-check and go for it.

But if you're sitting around with your cofounders trying to brainstorm a brilliant ploy and there are a half dozen ideas on the whiteboard and you all gradually decide, well, that one sounds best — you're probably forcing it. If you're giving out free food then no big deal, but if the word 'edgy' has crossed your mind, at least give yourself a couple of weeks to think things over before pulling the trigger.

And please: don't do stuff that's offensive. Don't put people or animals in harm's way. Think twice before making reporters look like fools. And don't trust yourself to make judgement calls on anything that might be 'pushing the envelope'.

* Deeply in debt with their site floundering, Airbnb's founders saw opportunity in the run-up to the 2008 Presidential Election — in the form of cereal. The small crew designed a matching pair of cereal boxes — Obama O's and Cap'n McCain's — got them printed on the cheap, and, after gluing them together by hand, sold them for $40 a pop. A whirlwind of coverage on CNN and myriad blogs later, they managed to sell over $30,000 worth of cereal, wiping out most of their debt. Later, as they tried to convince Paul Graham to let Airbnb into Y Combinator, it was the cereal stunt that sealed the deal. That's the story, at least...[5]

# Interviews

Aside from *Don't Lie*, the chief rule of any interview is to respect the reporter's time. If they ask you a question you don't want to answer, don't embark on a rambling soliloquy about your mission statement. You can dodge the question — people do it all the time — but keep it to a sentence or two. Another common tactic is to quickly segue and answer a different question that sounds vaguely related. Good reporters will call you out on this by getting annoyed and repeating the question, but some will just shrug and write down what you said.

Interview questions generally revolve around topics you'll find elsewhere in this book, but expect some curveballs. Good interviewers are always trying to get you to be more candid than you want to be.

Sometimes you'll get asked something you genuinely don't know the answer to. Don't feign ignorance, but if it's legitimate — say, a question around some metrics that you don't know off the top of your head — then it's okay to ask if you can follow up with an email.

Some reporters will read back your quotes to you if you ask them to (to check for accuracy and misstatements), but many will get annoyed if you even broach the subject. I'm in the latter camp: I'll read a quote back if there's something I want to double-check, but not on demand. But I'm a huge hypocrite: last time *I* was being interviewed, I asked the guy if he would run my quotes past me before he published (he said no*).

Also: remember what I said earlier about reporters being quoting machines. When you're talking on the record, particularly in the capacity of an interview, don't be careless with your words. Small gaffes make for big headlines.

---

* Based on the reporter's line of questioning I was concerned he might be taking a negative angle, so I cared more about making sure my quotes were accurate than I did about annoying him a little by asking to check them. We were discussing the always-contentious matter of paid video-chat consulting.

# An Unfortunate Surprise

It's 2009 and a dating startup is in the habit of buying up unprofitable-but-sizable Facebook applications, ripping out their code, and replacing it with their own. Suddenly that virtual aquarium you installed months ago is spamming you with invites to hook up with a woman who is doubtless unaware of your existence.

I am not a fan.

One afternoon I receive a tip that bolsters the case for karma: it seems that passwords on the dating site are now optional. Type in your email address, sans password, and you're swiftly carried to your main dashboard — or, anyone else's. I make an account, try logging in without password, and bam: full access. Yikes.

Standard ethical procedure around this kind of security hole is to contact the affected company before writing about it, so they can resolve the problem before my blog post sends twenty thousand people to go exploit it themselves. So, I call the CEO.

Our conversation is typical, as these things go. I tell him about the issue and ask him to please let me know when it's resolved. He agrees to do so, explaining that it must have been a bug they'd introduced that day in a code-push. He's taking it surprisingly in stride.

Sure enough, he calls me back within the hour to let me know that the bug has been fixed. He thanks me profusely. I say of course, and that the post will be up soon.

*Ughhhhhhhooooooooooooooooooooooooosh*

You know that scene in *Indiana Jones and the Last Crusade*, where the bad guy drinks from the wrong cup and his life-force gets sucked out by an omnipotent vacuum cleaner? That's what this sounds like, as every trace of happiness rushes out and despair fills the void.

"You're... you're going to write about this?"

I feel bad for the guy, really. Hate his company, wouldn't like him much if we met, but you witness a moment of reckoning like that you can't help but feel something. Maybe he won't have a job tomorrow.

I hit *Publish* ten minutes later.

I share this story to remind you that reporters are not your friends, or, even if they are, they have a deeply-held sense of duty to report the news. They are not in the business of doing you favors — especially when your service is running on sleaze.

Then again...

# Being a Source

*Warning: Here be dangerous waters. Being a source carries risk. You may be discovered. You may ravage your conscience and make reporters think you're a scumbag. Do not venture forth without a thorough understanding of what you're doing and why. This book probably isn't enough, but I want you to know about the system, because some of your competitors may use it against you.*

From anonymous tipsters to people close to the matter, they go by many names. Founders, employees, investors, even "friends". You may know a few yourself. They are sources, and they are the lifeblood of the press. They tell reporters secrets.

Reporters have romantic notions of meeting Deep Throat in a dank parking garage, but with rare exception these interactions are indistinguishable from the mundanity of office life. Sometimes, for the little stuff, the source just sends the goods in an email. Sometimes a vague email sets up a phone call (which leaves less of a record). Sometimes it's straight to phone call. Maybe there's newfangled ephemeral or encrypted messaging involved.

Say you shoot me a text. "Have a sec?"

"Calling you in 5," I write back.

We talk for a bit. Depending on your credentials and how you'll be attributed, I may have enough for a story. If your credibility is flimsy, or the publication is scrupulous, then I'll try to get a second source to verify. My heart is beating fast. Breaking news is a hell of a rush.

But why would anyone want to be a source?

Because it is how you turn the press into an ally, and wield it like a weapon. It's the loophole in journalistic ethics.

Reporters go to great lengths to absolve themselves of bias. Many shirk gifts, plaster disclosures everywhere, and try to hold themselves to a high moral standard. But by definition we are not required to share who our sources are with anyone but our editors (and, in some cases, we may even refuse to do that). A single source can change the trajectory of a reporter's career, and the readers never see a thing.

You ever notice how some reporters seem to tear into the same companies and executives, while treating others with a gentle touch? Maybe they just haven't gotten any good intel on the lucky ones — or maybe the lucky ones are giving them good intel. The way to a reporter's heart isn't through free gifts or anything else they'd have to disclose. It's whispering secrets in their ear.

This dynamic is intensely powerful, to the point that it makes the disclosures and ethics statements that reporters sprinkle across their sites feel reminiscent of the security theatre of the TSA. Sure, it lends a sense of professionalism and makes us feel better — but what exactly have we accomplished*? Favors beget favors, and the fear of losing a source can be enough for our minds to kill a story in the cradle, when it's just the wisp of an idea.

I digress.

No respectable reporter is going to engage in 'tit for tat' — give me a scoop, and I'll write something nice about you later. But reporters are human. The more we like you, the more likely we are to read your emails, to give you the benefit of the doubt when you need it. And, as the subjective nebula of thoughts swirls through our minds, we may like your product a bit better than we would have otherwise.

Not all sources are looking to curry favor with reporters. Some are aiming to undermine a competitor, others to change the dynamics of a negotiation (this is why you see so many stories leak about acquisition discussions). Reporters will always think about

---

* The danger with disclosures is not so much that they allow reporters to get away with an intentional agenda, but rather that they may lead reporters to believe that because they have gone through the motions, they have negated whatever biases they have.

What's worse, reporters almost always treat financial conflicts as if they are the predominant source of conflict, when really they're just the least awkward and easiest to pin down. Romantic relationships between reporters and members of the industry they cover are common (do I have to disclose a hook-up?), and many platonic friendships are laden in bias. Writing negative stories about friends is hard; the article may get written, but the words come out softer — and you'll rarely see a disclosure about it.

a source's motives, but the promise of a big scoop is often enough to outweigh the knowledge that we're instruments of corporate conniving.

Being a source is a dirty business, but it comes in shades. It doesn't necessarily entail selling out your friends and colleagues. Some sources traffic in publicly available information: they'll send a link to a website or two, perhaps with an additional line of context, and let the reporter connect the dots[†].

Some sources never preemptively email a reporter, but will pick up the phone and speak *Off the Record* or *On Background* when we call. Some will answer one question — "Will I look dumb if I print this?" — with a single word.

Please, think hard before doing any of this. And if you do decide to dabble, stick with the lighter end of the spectrum. Giving a reporter some *On Background* context about your competitor's shady tactics is miles away from leaking information that was given to you in confidence. The people you betray may never confront you, but it doesn't mean they don't know.

A few more thoughts:

**Don't say something you're not comfortable saying.**

You have no obligation to be a source. If a reporter calls you and asks something you don't want to answer — not even *Off the Record* — just decline to say anything. That's what most people do. The reporter might be annoyed, but it's par for the course. Just because you gave me a juicy tip a week ago doesn't mean you have to do it again.

---

† Sometimes people actually "pitch" like this: they'll send in a juicy tip about some secret project that no one is supposed to know about yet, but, really, they're in on it and are praying we'll bite.

**Become a go-to expert (but know what you're getting yourself into).**

If I'm in your shoes, I'm trying to establish myself as someone reporters turn to for *On Background* conversations, giving them context around breaking news and stories they're chasing without divulging secrets per se. This can help establish your credibility (and engender warm fuzzies) without getting your hands too dirty. But it's not for everyone — they'll still be trying to get you to share things you don't want to, and the stronger your relationship with them, the harder it gets to say no.

**Don't be wrong.**

If you're not really sure about something — maybe you heard a juicy rumor, but it's second-hand — then frame it that way. Feeding a reporter information that proves to be false (whether intentionally or not) is a good way to make them hate you, because it makes them look dumb.

**Just because you aren't named in the story doesn't mean people won't know it was you.**

We leave digital crumb trails everywhere these days. Leaking information securely is difficult. If you're doing something that could get you fired or worse, know what you're getting yourself into (note: you should not do research from any device or internet connection associated with your employer, and the NSA has done a good job reminding us that 'secure' often isn't).

And don't underestimate the 'human' risk — that someone will figure out who knows what, and decide that it was probably you who leaked it. They don't have to prove it to stop trusting you.

### Have a tip?

Most sites give you a way to submit an anonymous tip. Obviously this isn't going to curry favor with reporters (since they won't know you sent it), but it's a common way for competitors to try undermining each others' efforts. I guess whistleblowers use these too, but, mostly, they're magnets for spam, gossip, and takedowns.

### Read the warning again.

One PR person I spoke with was adamant that founders should not be sources because of the risks involved, though it happens all the time. Of course, you could always hire a PR person to do your dirty work...

# Never Should You Ever

A quick rundown on the many ways — some of which may seem benign — to get on a reporter's bad side. Let's start with the big one.

**Don't lie.**

If your mind instinctively whispered, "don't get *caught*", then fine. But know this: even if you aren't caught today, a year from now when a currently-happy-but-one-day-disgruntled-employee tips me off, I'm going to *crush* you.

Just don't do it, you'll sleep easier. And it's mostly easy to avoid: be vague, don't respond, or decline to comment if you don't have any other options.

"Next thing", you're thinking. But no — we're going to stay

here awhile longer, because I want you to understand that whatever you feel now is irrelevant. Your back isn't up against the wall, your ears aren't coated with the sticky wet breath of investors telling you to "do what it takes". It's all academic until you have six weeks of runway left and this launch is going to make or break your company and, really, will anyone get hurt if they don't know about this one little fudge on your part? There's no law against lying to a reporter…

But it's a bad idea. Because if I uncover your lie I'll cease to believe anything you tell me. I'll write about your falsity, my article will detail whatever other shady whispers I hear about you, and you'll be in no position to defend yourself because your credibility is shot. Other sites will pile on — scandal is always popular — and then Google will helpfully serve up reminders of your treachery to anyone who ever searches your name again.

Maybe the punishment outweighs the crime. But the press is a vindictive god, and lying is its cardinal sin.

Let's move on to some lesser transgressions.

**Don't suggest a headline (or copy).**

I am an independent thinker. I'll listen to your spiel and decide what I think is important and write my post as I see fit. Sure, I may quote a bunch of what you said verbatim, and then paraphrase the rest of pitch-points, but it's all coming from *me*.

The minute you say something that makes me think that you're trying to undermine this autonomy, I will stick my heels in the ground and do the opposite. It's offensive — outrageous, really — even if you were just trying to be helpful. Give me your easily-quotable pitch, lay it out on a silver platter, but don't dare *suggest* I print any of it.

**Don't act as if you are entitled to a story.**

You aren't, period. The day you give reporters the impression they are there to do your bidding is the day you start getting smeared and ignored.

This applies to tipsters, too. No matter how many solids you've done for me, don't say I owe you. You can complain a bit ("Come on man, this is really game-changing for us..."), but don't venture into "Dude, after that Google Fridge story I sent you..." territory.

**Don't use the wrong person's name or publication.**

No brainer. Happens all the time.

**Don't try to kill a story by pleading with a reporter.**

If reporters don't like getting told what to write in their headlines, we *really* don't like getting told not to write a story at all. If anything it'll make us feel like badasses (the bar isn't high), which doesn't bode well for you.

"But it's important," you say.

Don't. It'll probably backfire. But if you need to learn that lesson yourself...

You could explicitly try to trade a favor in return for killing the story — say, an exclusive on an upcoming announcement — but this is dangerously close to treading on my autonomy, which may destroy our relationship and lead to years of cynical coverage (or no coverage at all).

But sometimes there's some wiggle room.

Say I shoot you an email saying I heard you're working on a new product, and I'm going to be posting about it in twenty minutes. Your team has been pulling all-nighters for months to get ready for the big launch next week and here I am about to decimate their morale by writing a paper-thin post about "whispers"

I'm hearing — it's *news*, after all — and you *really* wish you could avert this.

So you ask if we can go *On Background* ("Ok.") and admit that yes, it's launching in a week and you haven't told anyone else about it.

"Bingo," I think to myself.

But then you say something I'm not expecting.

"Jason, I respect that it's your prerogative to report on this, but honestly it sounds like you're missing a big part of the story. If you'd consider holding off for now, I'd be happy to give you an exclusive on this and a full-briefing (interviews, videos, the works) when it launches next week."

And maybe I say "Well, if I heard about it, someone else might get tipped off, too." No dice.

But maybe my source was flimsy in the first place, and I'm worried that I really *am* missing a big piece of the action. So I bluff hard about how much I know and finally cave. Success[*].

### Don't redefine the word "No".

During your exchanges with reporters, you'll notice they have a tendency to simply not respond — not even a simple "thanks, but no thanks." Partly because they can get away with it, partly because they intend to get back to you later but forget, and mostly because they are routinely punished for doing the courteous thing and saying, "No".

Why? Because PR people love to interpret "No" as, "I don't like this pitch, but tweak it a bit and maybe I will!"

---

[*] This, of course, assumes that the carrot you're dangling has some heft to it. If I decide you've misled me, I'll exact revenge (and: in the process of trying to convince me not to publish, you may have confirmed something I was only half-sure about). Dangle with caution.

So if we *do* take the time to say "No.", then that's that.

**Don't ask to see a story before it hits the presses.**

Some people ask to see or have their stories read back to them before they go live (presumably with the intention of correcting any errors). Reporters *hate* when people ask about this — autonomy and all that — and usually it's only the minor leaguers who do.

That said, some magazines and even a few rogue newspaper reporters are happy to do this[†]. So don't ask under any circumstances, unless you're pretty sure they're cool with it.

Journalism, folks.

**Don't pull a fast one (unless you're okay with the consequences).**

Say I call you out of the blue and ask for confirmation that you've just landed a major partnership. You stall — it's a story you've been prepping for weeks, pre-briefing other reporters and everything — and you *really* don't want me to be the one to do it. So you tell me it's not true, or that you'll get back to me in half an hour, and then you run off and give the green light to the reporter you like better.

Sure, you can do it, but I'll hate you for it. Which brings us to...

---

† "However, readbacks are common at magazines such as Time, National Geographic, and The New Yorker. Some newspaper reporters also make it a standard practice. Jay Matthews of The Washington Post said he has read or faxed copies of stories to sources for the past 10 years." - Ron Smith, *Ethics in Journalism*[4]

# Vengeance

Remember when I said I wrote this book partly out of revenge? Not a healthy emotion, but one that reporters are especially well-positioned to execute on. I won't *say* anything to you about it, of course — it'd be terribly unprofessional and leaves me open to accusations of compromising my journalistic integrity. But that's the beauty of being a reporter: I don't have to say a word to rain on your parade.

Instead, I'll just ignore everything you ever pitch me, and I'll share your exploits with my coworkers so that they do the same. And then when there *is* something about you to potentially go negative on, I'll really sink my teeth in, badgering you with questions and writing a piece that makes you sound like a fool (complemented by an old, unflattering photo of you looking like a fool).

I'm not saying that reporters keep a blacklist, or even if we did, that you'd be on it. It's more subtle than that. Ask me why I don't like one company over another and I'll point to a lack of innovation, but probably the embargo they screwed up six months ago has something to do with it, and the fact that they gave their big exclusive to my biggest rival (hack).

I've held grudges. I've done solids. I've always tried to do the right thing, but there is no right thing when I'm pulling subjective ideas out of my brain — and the sweeter that stew when I think of you, the better.

If you take away one thing from the book, it should be this: reporters are just people.

It's why everyone sucks up to us constantly, with free food and drinks and swanky events.

It's why sources tell us secrets, why there's always someone asking if we have everything we need, why executives pick up the phone at all hours.

It's why people are always trying to make us feel special.

You don't need to be on our good side, but boy, does it help.

# The Bad News

When I set out to write this book my thesis was this:

> *If you send a great pitch to a reporter, it doesn't matter whether they know you, or if you have an intro. A well-crafted pitch from a founder tops all that stuff (or at least, puts you on the same footing).*

I was naive and foolish.

Remember: I'm not currently a tech reporter. Much of this book was written after I left *TechCrunch,* and as part of my research I've done some startup PR consulting (also: it pays well). The notion of doing 'real' PR — leveraging my friendships with reporters and pitching them directly — makes me queasy, so I set a rule for myself and my clients: I would not interact with reporters on their behalf, opting instead to lurk in the shadows and help with pitch-writing and strategy.

I figured if the pitch was great, perfectly-formed and no bull-shit, sent by the founder to just the right person, then reporters would take notice on their own accord and I wouldn't need to get my hands dirty. Wha-bam! Press.

It hasn't worked that way. Twice, out of not-so-many tries, I've found myself in the tense PR purgatory of waiting for a reporter to respond — to even *acknowledge* the pitch — justification of my hefty fee hanging in the balance.

"Should I try pinging him again?" the founder asks.

"Give them til tomorrow," I say.

"Any word?" me, the next day.

"No. Shooting over the follow-up now."

Another twenty-four hours pass.

"Anything?"

"Nope :-/"

And then after much hand-wringing I'd bend my rule, driven in small part by a sense of justice, large part a sense of ego.

Both times, an email to a former colleague did the trick. I disclosed I was being paid, but that I thought it was a solid story (I wouldn't have accepted the business in the first place if it weren't, I rationalized).

Christ, I felt dirty.

Maybe there are reliable press tactics that obviate the need for relationships. Obviously I don't know them (I'm all ears). But if you're only as good as me, relationships will account for a big chunk of your batting average. If you're cold-pitching and fiz-zling, this is suspect number one.

And if you're having a hard time building those relationships, maybe your team could use someone who's better at it. Just make sure your smooth operator comes with some moral fiber.

# The Newsroom

Conventional wisdom dictates that reporters are driven to sensationalize, libel, and otherwise exploit the excitement pumps in readers' brains because of our insatiable thirst for page views. Page views generate revenue for Mother Blog, which in turn lead to bonuses, praise, and career advancement. And so we have this relentless, cynical race to cover anything and everything that can be passed off as "news", where nuance and accuracy are regularly trampled.

Many reporters would agree with this assessment. Yet we are just as likely to feel as if we are an inconsequential part of the problem, because we know how rarely we're actually thinking about maximizing traffic. The honest truth is that in our heart of hearts we want to write stories that have the most importance, the most *impact* — for which page views are merely a proxy.

And somewhere deep in our minds, The Devil (as voiced by James Earl Jones) cackles with glee.

The problem with the Page Views narrative is that it gives reporters too much credit. We are too busy trying to get the story looking decent enough to hit *Publish* to have time to consider why we're choosing one headline over another. It's just that certain stories and phrases *feel* right; our fingers are guided by forces subtle and mysterious, the sort that bypass and manipulate logic by harnessing the potent powers of emotion.

That's where the booze of validation comes in — that craving for the serotonin spikes of comments and Likes and views — and indeed, it accounts for many a sensationalized and poorly-considered post. But these boozy impulses are short-term. It is only over extended exposure to them and the culture they foster that our judgement suffers long-term damage: a cirrhosis of the conscience, if you will.

The crucible for these issues is the newsroom.

The best thing about the newsroom is that it is a fantastic cure for loneliness. Because while its previous incarnations were restricted to the confines of a physical office, the modern newsroom is available on iOS, Android, and the web. It is with us at all times, our thumbs trained to whisk us there when we feel bored or alone or we think too hard about how the news cycle feels like a hamster wheel, how even our successes are forgotten in a matter of hours.

But back inside the newsroom we are relieved to remember that we are part of a team, a powerful team capable of toppling Power. Of chasing Truth. And of wickedly clever banter. What fun we have, our running commentary of the day's news and idiots and gaffes. It can be brutal, sure — nothing like having your boss slam your post in front of the entire company — but we're part of something, in a way that satisfies our base human desire for belonging.

And it *was* a bad post.

At its best the newsroom is a bastion of camaraderie: ideas for angles, offers to contact each other's sources, and a deep institutional knowledge worth countless hours of research. But these kind words belie an undercurrent of anxiety and jealously.

"Nice headline," an editor says to a colleague, and the rest of us rush to agree; few things make reporters swell with such pride. Yet as our punny peer floats on Cloud 9, the rest of us are quietly stewing, wondering why no one said anything about *our* cheeky turns of phrase.

Take the negative feelings evoked by a poorly-performing and utterly inconsequential Facebook status update and magnify them with the knowledge that how you fare on *this* feed may dictate your career path and you have a sense of how deep the tensions flow. You're only on the same team until one of you gets noticed by *The New York Times*, or a VC firm, or whatever escape hatch you're reaching for. No one wants to blog forever.

Naturally the flames of competition spur an intense drive to improve, to one-up, to become a bigger name than the others. Be faster. Cleverer. Don't second guess yourself so much. Management stokes these flames by distributing traffic reports under the guise of helping us discern "what works", as if it weren't already apparent from 'social' score counts splayed across each post.

And so we find each writer playing their own half-conscious game of chicken. Is this headline too cruel? *Nah, it's not nearly as bad as the one* he *wrote last week.* Is my source concrete enough to publish this? *Well, if I just hedge things a bit...*

The exploits of the competition compound these effects — new tactics spread like a disease from site to site — but the competition are a bunch of hacks. It isn't until bad behavior is condoned and encouraged at home, by the people we know to be decent and friendly, that we manage to convince ourselves that we are doing nothing wrong.

Eventually, thoughtlessly, we find ourselves impassioned with new convictions, a trigger-happy sense of outrage, and a numbness to the pain we inflict on others. Occasionally we worry we have gone too far and we look to our teammates, but they are there doing the same, are even further past the edge, looking back and cheering us on. We're all marinating in the booze.

# The Callous

We all have those memories our brains like to play back, unsolicited. One of mine that's been in heavy rotation...

There are five of us, the faces blurry but the cast was small so they aren't hard to guess. We're sitting in my boss, Michael Arrington's house, in the hallway we call an office. I just finished screaming into the phone, chewing out a hapless PR person over something too trivial to remember.

"Jason used to be nice," Sophie says. And everyone laughs.

For a long time I convinced myself she was joking.

You start off a little numb. It's a game: the scoops, the head-lines, the sources — all a frenzied race to beat the opposition, the companies mere abstractions strewn across a playing field encir-cled by an endless array of scoreboards.

In time a small voice grows louder. Those capitalized words represent people, you know. Each critical barb, each cynical quip, absorbed by their friends and lovers, parents and children. Why not soften the harsh words, so that perhaps they won't be seared into so many minds?

But the inbound torrents are relentless. How many hours does each plea represent? Hundreds? Thousands? This one you met at a party, apparently. How many times have they relived that conversation? "I'd love to hear more about it," you probably said. Here's a card. Now leave me in peace.

The masses crave dark times, to see the powerful crumble. They are sympathetic to neither the subjects of your articles nor the pressures you face. "What fools," they say to the newcomers. "Good riddance," they say to the fallen. Even the good news is poisoned with reminders of your bias and ineptitude.

In this caustic environment survival is a matter of adaptation. A callous grows. It is not the naive numbness as before, but a selective desensitization, a set of internal rules that allow you to press on.

*Pay no attention to the biting criticism of strangers. Their words stem from jealousy, ignorance, and hate.*

*Bring justice to those who deserve it. They knew the stakes.*

*Do not dwell on the despair of the unanswered founder. There are countless others who have sacrificed just as much.*

I thought I knew where the callous started and ended, that I was in control.

I became someone else.

# Burnout

I don't know when it happened, exactly. It was probably a gradual thing, and it wasn't til I got out that I realized how deep in I was. But some moments I look back on and think, "maybe it was then."

A startup is throwing an event for some new feature. Not Facebook or Google, where even minor announcements are arguably newsy because they impact a lot of people. Just a startup with a lot of money and not a little desperation.

A smallish pack of reporters is here, enough to fill half of the plastic chairs laid out in our honor. The CEO comes out, there are slides; big ambitions with tenuous ties to an extant product. It's taking forever and — by the way — the embargo lifts in thirty minutes.

The guy may as well told us to go fuck ourselves. It's bad enough we just burned our afternoons on a mediocre event (and traveled here for the pleasure). Now we're under the gun of an obnoxious and arbitrary embargo, no time to deal with the smoldering fires in our inboxes that wouldn't be there if we'd skipped the damn thing. Also, I'm hungry.

We hunch in silence — not even a table to work off — to crank out the stories.

*Rat-tat-tat-tat.* Pause, sigh. *Rat-tat-tat-tat.*

You tune things out when you're in the zone. Everything becomes peripheral vision, even the words you're writing. Once you're there you stay as long as you can.

Paragraph one, *Today at an event in San Francisco...*

Now stats: *copy, paste, revise.*

Next graf —

"...SOME ADDITIONAL RESOURCES AND SCREEN-SHOTS."

Ugh.

The zone is gone, I'm back here, she's waving a thumbdrive at me.

"Thanks."

Trying to get back in. It's like trying to get a model train on the tracks, you've gotta go by feel. Rereading what I've got, searching for the flow —

"...HAVE EVERYTHING YOU NEED?"

"I hate you."

It's like I punched her. Faster than a punch, I don't even have a chance to stop mid-way. It's out there, it smacks her hard in the gut.

My mind's buzzing, I shouldn't have said that, I feel awful.

"I'm sorry," I say.

She says it's okay.

I finish the post, not caring about it any more. Someone else can have the headline.

I find her, apologize again.

"It's okay," she says.

It isn't.

# The End of the Universe

A lot of startups are doomed. Most of them. Sometimes they fail because they can't secure that one thing they need, and sometimes they never had much chance.

One reason the press is so attractive is that it serves as tangible validation. There are other ways to get that, but press may feel the most attainable. Which is why it's devastating when not a single reporter seems to give a damn.

Maybe it's them. As I've said ad nauseam, a lot of this comes down to luck and relationships and things you don't have much control over.

Maybe it's me, because no book like this could be perfect, and I'm certainly not.

Or maybe it's you. The company, at least. Maybe your startup just isn't going to cut it.

There's no litmus test to tell you which is true. Fact is, there are successful startups that get ignored by the press for ages until they get noticed.

But if you try to get coverage for your launch and that fizzles, it'd be wise to question your assumptions. If the next go-around fizzles again, question them harder. Go to an event and find a reporter or investor — someone good, someone who cares — and preface your pitch by saying, "I'm not looking for coverage, but I'd love your feedback on why this might not be interesting." Maybe they'll give you a listen.

And again, you don't *need* press. Keep cranking and maybe some day the bloggers will have to play catch-up. But crank long enough without any sparks and at some point you need to look at the signs and figure out whether they're starting to coalesce. It wouldn't mean what you built was bad — many fantastic products never get the attention they deserve. Just as likely, you had unlucky timing that hinged on events no one could've predicted. So much of this is luck.

It'd be a failure. I could say it wouldn't — that you've learned so much that you've won — but we both know that isn't true. You're putting it all on the line, if coming home empty-handed isn't the definition of failure then what is?

But let's give you some credit.

You had an idea you needed to see through, and instead of shrugging your shoulders and putting it on the back-burner til it dies like everyone else does, you went for it. Most people are too scared.

And when you second-guess yourself, remember: every

founder steers their company through countless decisions. Often, the events that prove most important are heavily influenced by luck. There are things you could have done differently, but the stuff that seems obvious in hindsight might not have worked out anyway — and you have no way of knowing which decisions were important to begin with (it's the little things that snowball).

Sorry, I'm being preachier than usual. But spare me time for one more thought.

I've met a lot of people you'd probably call successful, some of them fabulously so. They are rich, they are brilliant, some are famous. A few have arguably met that ridiculously lofty goal of putting a dent in the universe.

I don't spend my days wishing I were them. Few of them seemed that happy. They were still looking for the next rung. There's always a next rung.

The people I envy are the ones who are always surrounded by good friends. Who can make it through a whole sunset without feeling the itch to annotate and share it. Who laugh like kids on summer break.

The universe doesn't give a damn about any of us.

Try to find who does.

## About the Author

Jason Kincaid is a writer. He tries a lot of other stuff too, and hopes that one day the first sentence in his hitherto (and justifiably) nonexistent Wikipedia entry will be pretty long.

He joined *TechCrunch* in 2008, where he wrote thousands of articles — including several of consequence — and interviewed many of the top figures in the tech world. He has emceed numerous marquee events and appeared as himself on HBO's hit comedy, *Silicon Valley*.

For his latest essays, music, and more visit: **jasonkincaid.net**.

# Additional Content

I've covered as many aspects of dealing with the press as I can think of, but there are doubtless others that should be included. In light of this, I've set up a website at *BurnoutPR.com*, where I will post updates to this book for free (I'll eventually add these to the ebook as well, whose updates will be free).

This book originally contained chapters on subject lines and what to include on your homepage. They're boring so I took them out, but they're also at *BurnoutPR.com*. While you're here: put a *Press* link at the bottom of your site with logos, screenshots, and a contact email.

Other people have written at length about some of these subjects. I keep a list of some favorites on the site.

If there are any additional topics that you think would be broadly useful (or anything in this book you find particularly misguided) please let me know.

# Acknowledgements

Thank you to Matt Dawson for his fantastic cover artwork and for being a true gentleman.

Thank you to Ed McManus, Mark Hendrickson, Sarah Passe, and Andrew Fried for reading early versions of this book (and treating my ego with a gentle touch). And to Andrew Kovacs and the other PR people close to the matter who spoke on the condition of anonymity, whose feedback proved invaluable.

Thank you to my friends and family for their encouragement and for only occasionally rolling their eyes when I stopped talking mid-sentence to record a voice memo relevant to this book on my phone.

The unicorn illustrations are by *Lindybug* (used under license).

The style of the graph in *Why Not to Get a PR Person* was inspired by *xkcd*, and was partially generated using Mathematica along with code found on StackExchange. Thanks to Serkan Piantino for his assistance with the asymptotes.

Much of this book was written with *Tycho's* albums *Dive* and *Awake* on repeat.

And to my fellow *TechCrunch*ers: couldn't have picked a better crew to burn out with. Who's up for Chipotle?

# Works Cited

*Orthodontists Earn More Than CEOs: 10 Facts About Wages.* Eric Morath. April 1, 2014. *The Wall Street Journal.* Web, accessed April 2014.

*On Language; Off the Record.* William Safire. October 29, 1989. *The New York Times.* Web, accessed May 2014.

*For the Record, What "Off the Record Means".* Chatterbox. June 22, 1999. *Slate.* Web, accessed April 2014.

*Ethics in Journalism, 6th Edition.* Ron F. Smith. Wiley. Published 2008. Kindle edition, locations 1744-1746.

*Brian Chesky - Founder of Airbnb at (Y Combinator) Startup School 2010.* Ernest Semerda. October 22, 2010. YouTube, accessed April 2014.

# Further Reading

Rafe Needleman has been posting quippy (and funny) PR tips on his site, Pro PR Tips (http://www.proprtips.com) for years; you can get them in book format, too.

Ed Zitron, a PR person based in San Francisco, wrote a book called *This is How You Pitch*. It is targeted toward newbie PR people (versus entrepreneurs) and has some good anecdotes. I don't agree with everything he says, but who says I'm right about everything?

You'll find a more extensive list of further reading at
BurnoutPR.com

This book will change a bit in the future, so I may as well include a version number.

This is version 1.0.4

---

1.0.4 - *Refactored cover, more unicorns. Oct 2017.*

1.0.3 - *Enhanced glyphs. Aug 2015.*

1.0.2 - *Design tweaks, bug fixes. Nov 2014.*

1.0.1 - *Bug fixes. Oct 2014.*

1.0  - *Initial Release. Sept 2014.*

23912943R10117

Made in the USA
San Bernardino, CA
31 January 2019